http://www.chelt.ac.uk/gdn

Teaching and Learning Geography with Information and Communication Technologies

Ifan Shepherd
Advisor: Peter Newby

Middlesex University

Series edited by Phil Gravestock and Mick Healey
Cheltenham & Gloucester College of Higher Education

Published by:

Geography Discipline Network (GDN)
Cheltenham & Gloucester College of Higher Education
Francis Close Hall
Swindon Road
Cheltenham
Gloucestershire, UK
GL50 4AZ

Teaching and Learning Geography with Information and Communication Technologies

ISBN: 1 86174 030 1 ✓
ISSN: 1 86174 023 9

Typeset by Phil Gravestock

Cover design by Kathryn Sharp

Printed by:

Frontier Print and Design Ltd.
Pickwick House
Chosen View Road
Cheltenham
Gloucestershire, UK

Teaching and Learning Geography with Information and Communication Technologies

Geography Discipline Network (GDN)

Higher Education Funding Council for England
Fund for the Development of Teaching and Learning

Dissemination of Good Teaching, Learning and Assessment Practices in Geography

Aims and Outputs

The project's aim has been to identify and disseminate good practice in the teaching, learning and assessment of geography at undergraduate and taught postgraduate levels in higher education institutions.

Ten guides have been produced covering a range of methods of delivering and assessing teaching and learning:

- Teaching and Learning Issues and Managing Educational Change in Geography
- Lecturing in Geography
- Small-group Teaching in Geography
- Practicals and Laboratory Work in Geography
- Fieldwork and Dissertations in Geography
- Resource-based Learning in Geography
- Teaching and Learning Geography with Information & Communication Technologies
- Transferable Skills and Work-based Learning in Geography
- Assessment in Geography
- Curriculum Design in Geography

A resource database of effective teaching, learning and assessment practice is available on the World Wide Web, http://www.chelt.ac.uk/gdn, which contains national and international contributions. Further examples of effective practice are invited; details regarding the format of contributions are available on the Web pages. Examples should be sent to the Project Director.

Project Team

Lead site: *Cheltenham & Gloucester College of Higher Education*
Professor Mick Healey; Dr Phil Gravestock; Dr Jacky Birnie; Dr Kris Mason O'Connor

Consortium: *Lancaster University*
Dr Gordon Clark; Terry Wareham
Middlesex University
Ifan Shepherd; Professor Peter Newby
Nene — University College Northampton
Dr Ian Livingstone; Professor Hugh Matthews; Andrew Castley
Oxford Brookes University
Dr Judy Chance; Professor Alan Jenkins
Roehampton Institute London
Professor Vince Gardiner; Vaneeta D'Andrea; Shân Wareing
University College London
Dr Clive Agnew; Professor Lewis Elton
University of Manchester
Professor Michael Bradford; Catherine O'Connell
University of Plymouth
Dr Brian Chalkley; June Harwood

Advisors: Professor Graham Gibbs (*Open University, Milton Keynes*)
Professor Susan Hanson (*Clark University, USA*)
Dr Iain Hay (*Flinders University, Australia*)
Geoff Robinson (*CTI Centre for Geography, Geology and Meteorology, Leicester*)
Professor David Unwin (*Birkbeck College, London*)
Dr John Wakeford (*Lancaster University*)

Further Information

Professor Mick Healey, Project Director   Tel: +44 (0)1242 543364   Email: mhealey@chelt.ac.uk
Dr Phil Gravestock, Project Officer   Tel: +44 (0)1242 543368   Email: pgstock@chelt.ac.uk
Cheltenham & Gloucester College of Higher Education
Francis Close Hall, Swindon Road, Cheltenham, GL50 4AZ, UK   [Fax: +44 (0)1242 532997]

Contents

Editors' preface

This Guide is one of a series of ten produced by the Geography Discipline Network (GDN) as part of a Higher Education Funding Council for England (HEFCE) and Department of Education for Northern Ireland (DENI) Fund for the Development of Teaching and Learning (FDTL) project. The aim of the project is to disseminate good teaching, learning and assessment practices in geography at undergraduate and taught postgraduate levels in higher education institutions.

The Guides have been written primarily for lecturers and instructors of geography and related disciplines in higher education and for educational developers who work with staff and faculty in these disciplines. For a list of the other titles in this series see the information at the beginning of this Guide. Most of the issues discussed are also relevant for teachers in further education and sixth-form colleges in the UK and upper level high school teachers in other countries. A workshop has been designed to go with each of the Guides, except for the first one which provides an overview of the main teaching and learning issues facing geographers and ways of managing educational change. For details of the workshops please contact one of us. The Guides have been designed to be used independently of the workshops.

The GDN Team for this project consists of a group of geography specialists and educational developers from nine old and new universities and colleges (see list at the front of this Guide). Each Guide has been written by one of the institutional teams, usually consisting of a geographer and an educational developer. The teams planned the outline content of the Guides and these were discussed in two workshops. It was agreed that each Guide would contain an overview of good practice for the particular application, case studies including contact names and addresses, and a guide to references and resources. Moreover it was agreed that they would be written in a user-friendly style and structured so that busy lecturers could dip into them to find information and examples relevant to their needs. Within these guidelines the authors were deliberately given the freedom to develop their Guides in their own way. Each of the Guides was refereed by at least four people, including members of the Advisory Panel.

The enthusiasm of some of the authors meant that some Guides developed a life of their own and the final versions were longer than was first planned. Our view is that the material is of a high quality and that the Guides are improved by the additional content. So we saw no point in asking the authors to make major cuts for the sake of uniformity. Equally it is important that the authors of the other Guides are not criticised for keeping within the original recommended length!

Although the project's focus is primarily about disseminating good practice within the UK a deliberate attempt has been made to include examples from other countries, particularly North America and Australasia, and to write the Guides in a way which is relevant to geography staff and faculty in other countries. Some terms in common use in the UK may not be immediately apparent in other countries. For example, in North America for 'lecturer' read 'instructor' or 'professor'; for 'staff' or 'tutor' read 'faculty'; for 'postgraduate' read 'graduate'; and for 'Head of Department' read 'Department Chair'. A 'dissertation' in the

UK refers to a final year undergraduate piece of independent research work, often thought of as the most significant piece of work the students undertake; we use 'thesis' for the Masters/PhD level piece of work rather than 'dissertation' which is used in North America.

In addition to the Guides and workshops a database of good practice has been established on the World Wide Web (http://www.chelt.ac.uk/gdn). This is a developing international resource to which you are invited to contribute your own examples of interesting teaching, learning and assessment practices which are potentially transferable to other institutions. The resource database has been selected for *The Scout Report for Social Sciences*, which is funded by the National Science Foundation in the United States, and aims to identify only the best Internet resources in the world. The project's Web pages also provide an index and abstracts for the *Journal of Geography in Higher Education*. The full text of several geography educational papers and books are also included.

Running a consortium project involves a large number of people. We would particularly like to thank our many colleagues who provided details of their teaching, learning and assessment practices, many of which appear in the Guides or on the GDN database. We would also like to thank, the Project Advisers, the FDTL Co-ordinators and HEFCE FDTL staff, the leaders of the other FDTL projects, and the staff at Cheltenham and Gloucester College of Higher Education for all their help and advice. We gratefully acknowledge the support of the Conference of Heads of Geography Departments in Higher Education Institutions, the Royal Geographical Society (with the Institute of British Geographers), the Higher Education Study Group and the *Journal of Geography in Higher Education*. Finally we would like to thank the other members of the Project Team, without them this project would not have been possible. Working with them on this project has been one of the highlights of our professional careers.

Phil Gravestock and Mick Healey

Cheltenham

July 1998

All World Wide Web links quoted in this Guide were checked in July 1998. An up-to-date set of hyperlinks is available on the Geography Discipline Network Web pages at:

http://www.chelt.ac.uk/gdn

About the author

Ifan Shepherd

I am a principal lecturer at Middlesex University, where I teach courses in geodemographics, GIS, remote sensing, and Internet marketing. For many years a member of the School of Geography and Environmental Management, I have recently joined the Business School where I run the MUBS Online project. This aims to co-ordinate the use of communication technologies to provide effective support for all School activities.

My current research interests include data visualisation (including the use of virtual reality software), information quality on the Internet, the mapping of urban historical data, and the conceptualisation of skills. I have been active in the field of educational research for most of my working life, have published widely on computer assisted learning in geography, and am a founding member, former editor and continuing editorial board member of the *Journal of Geography in Higher Education*. I am currently involved in several collaborative research projects, including one aimed at creating spatial databases from the original data gathered by Charles Booth for his famous social surveys of late-Victorian London.

i   What this Guide is all about

"The successful exploitation of C&IT [Communications and Information Technology] is pivotal to the success and health of higher education in the future."

(NCIHE, 1997, para 13.57)

The computer is without doubt the most hyped educational technology of recent years. And yet the terminology surrounding its educational uses is often impenetrable (hence our Glossary), its benefits to teaching and learning are hotly disputed, and in some quarters (for example, the schools system in North America) there are signs of a backlash against lavish spending on IT in education. Against this background, the primary aim of this Guide is to examine how, when and where information and communication technologies (ICT) can be used to advantage in the teaching and learning of geography.

This Guide is not meant to be encyclopaedic. Its contents are highly selective, and focus on providing advice and examples from other practitioners that will allow you to make effective use of educational technology in your own courses. It might therefore be useful to begin by spelling out what the Guide will and will not cover.

What you will find in this Guide

- More than just IT. Rather, it is concerned with the educational use of both information and communication technologies in geography, often in combination. Computers alone are no longer the sole driving force of technology-mediated learning; arguably, communication technologies will become of far greater importance during the next few years. This combination has been recognised by the Dearing Committee of Inquiry into Higher Education, who refer throughout their Report (NCIHE, 1997) to 'Communications and Information Technology' (or C&IT), and also by HM Government's proposals for a National Grid for Learning (DfEE, 1997) which refers to 'Information and Communication Technologies' (or ICT). Elsewhere in the world, the term telematics is also widely used, with much the same meaning. We adopt the shorthand 'ICT' throughout this Guide, because it is becoming the more widely used phrase.

- The use of technology to support and enrich what you do, not necessarily to replace what you do. Basing an entire geography curriculum on ICT may indeed be what some geography departments decide to do, but wholly technology-mediated courses are not the focus of this Guide. Our intention is not to provide a blueprint for how silicon can replace flesh and blood, but to indicate how ICT can be used to good effect in various parts of the curriculum, for and with both teachers and students.

- Learning as well as teaching. As will become clear shortly, the Guide is primarily about the learning opportunities afforded by ICT to students of geography. Nevertheless, it will not entirely neglect the more obvious benefits of ICT for tutors, and will also indicate some of the changes in classroom practice that tutors may need to make in order to reap the fullest benefits from using ICT technologies.

- The experiences of both geographers and non-geographers. The primary intention of this and other Guides in the series is to share best practice amongst geographers in higher education in the UK. However, because little has been published on the uses made of ICT by geography teachers, we will cast our net beyond geography and include examples of good practice from neighbouring disciplines and also from mainstream educational technology where appropriate.

- Warts and all. Although this Guide concentrates on reporting best practice, it would not be doing its job properly if it ignored, or hid from view, some of the 'rough edges' of ICT. We will therefore be both critical and cautionary wherever the subject demands it.

What you won't find in this Guide

Above all, this Guide is meant to provide practical help in adopting ICT support for situations you consider important. For that reason, it will not contain:

- A theoretical discussion of educational technology principles or an abstract classification of educational software — these can be found in some of the references included in the Resources Database (see Section iii).

And because technical details about ICT get rapidly out of date, the Guide will not contain:

- Lists and technical descriptions of items of current educational software — these can be found in various catalogues (see Section 5.3).

- Reviews of specific ICT resources — these are available in a variety of existing sources (see Section 5.3).

- Specifications of the hardware needed to implement ICT-based activities — these change almost by the month, and would rapidly make this section of the Guide obsolete.

Finally, it is important to mention the close relationships between this Guide and several others in the series. Because the use of computer and communication technologies touches so many aspects of the curriculum, and can be used in so many ways to support teaching and learning in geography, it is hardly surprising that a fair degree of overlap will exist between the contents of this Guide and particularly those concerned with practical work, field work, transferable skills and assessment (Birnie & Mason O'Connor, 1998; Livingstone *et al.*, 1998; Chalkley & Harwood, 1998; Bradford & O'Connell, 1998). Also, if you are looking for ideas on how to use more conventional educational technologies such as video or readers, then take a look at the companion guide on 'Resource-based Learning in Geography' (Healey, 1998). It is hoped that the treatment in each of these Guides will be complementary to one another. Cross-references will be provided wherever relevant.

ii   Using this Guide

Before launching into the meat of this Guide, we need to say a word or two about how to use it and the related resources.

The main body of the Guide is divided into three sections. The first section discusses some of the key issues in the application of ICT to geography teaching, adopting the FAQ ('Frequently

Asked Questions') format widely adopted on the World Wide Web. The second section will review a range of problems that may be solved and opportunities that are opened up by using ICT. The final section considers the characteristics and educational potential of three technologies that are currently 'hot topics' in the educational world: multimedia, the Internet and the World Wide Web. With so much (dis)information and marketing hype surrounding these technologies, it is often difficult to form a clear view of what they genuinely have to offer in the classroom, and to geography in particular. This final section therefore attempts to get to the heart of these technologies, and attempts a preliminary assessment of their educational strengths and weaknesses.

Throughout the Guide, we refer to literature that provides practical guidance on using ICT in geography. Surprisingly, perhaps, relatively little has been published by geographers in higher education on their use of educational technology; most of the experiments reported in print appear to come from courseware developers. As a corrective to this bias, most of the case studies included in this Guide will be drawn directly from 'end users', for instance from teachers whose innovations may not have appeared in print. Reference is also made throughout the text to relevant information located on the Internet. The addresses (or URLs) of this material are gathered together in a sorted list at the end of the Guide; in-text references consist of an easily recognised acronym for the material, for example [GISCC].

Because this Guide has been designed to be used alongside the accompanying Resources Database (Section iii), you are strongly encouraged to:

- Read the thumbnail case studies in this Guide that are most relevant to your particular needs, then explore the fuller case studies in the Database for those applications that interest you most.

- Contact the innovators directly to discuss how their experience can be transferred to your own particular courses and institution.

- Follow up references to the published literature on appropriate applications of IT.

- Acquire resources indicated in the Database where they are relevant to your needs.

If you want a general review of the contributions ICT can make to geography teaching and learning, then a useful starting point is the chapter on 'Computer Assisted Learning and Teaching' in Gold *et al.* (1991). This can be supplemented by reading the numerous papers on the subject that have appeared in the *Journal of Geography in Higher Education*; most of those up to 1990 have been re-issued as a special publication (Unwin & Wood, 1990). Finally, a complementary review of successful applications of ICT in American higher education is provided by Boettcher (1992).

iii   The Resources Database

What is the resources database?

Accompanying this Guide is a database of resources concerning 'good practice' in geography teaching. This provides a common and unified resource for all 10 GDN Guides, and includes three types of material:

1. Case study abstracts

These are abstracts of interesting teaching, learning or assessment practices in geography. Each of the case study abstracts is accompanied by the name and contact details of the innovator who produced them, so you can contact them directly for further information.

2. Guide to the literature

Again, there are three sets of bibliographic resources in the database, all of them searchable:

- A complete listing of all papers published in the *Journal of Geography in Higher Education*. This was originally prepared by David Riley, who has kindly allowed it to be adapted for use in its current form. Abstracts for all papers have been added.

- An annotated bibliography of the use of CAL in geography. (This was originally prepared for a special CTI publication by Ifan Shepherd, and was first published in 1992.)

- A complete list of the references and World Wide Web links included in all 10 of the GDN Guides.

3. Other Resources

These include two broad types of material:

- Resource materials for use in teaching and learning. Some of these are geography specific, others are more generic.

- Links to Internet sources — for example, URLs, list servers, email addresses.

Why is the Resources Database not printed?

The Resources Database has been produced in computer-readable form for the following reasons:

- to enable it to be made available in several forms: as a distributed CD-ROM, or as pages on the Internet;

- to enable the contents to be rapidly searched by computer software;

- to enable the contents to be extensively cross-referenced (using embedded 'links');

- to enable it to be easily updated, including by users (through its moderators);

- to reduce reproduction costs.

The case study material in the Database has been specially designed to be read on screen. (For the technically minded, the contents have been formatted as HTML documents.) However, it is recognised that you may prefer to read from a paper copy, so we have provided facilities for printing out any of the material included in the Database. You are welcome to print any of the contents if you so wish.

How do I use the Resources Database?

You can use the Resources Database in two ways:

- Indirectly, through guidance provided in the individual GDN Guides. Each Guide makes extensive reference to the materials in the Database — you can think of each of them as providing a 'guided tour' of the contents of the database, with each tour addressing a particular aspect of teaching and learning.

- Directly. The Database can be browsed on screen, section by section, much like leafing through a printed book.

Because the Database is extensively cross-referenced, you will be able to follow 'links' that relate individual case studies to references and other resources within the Database. There are also copious cross-links between the Database and the individual Guides.

1 Teaching and learning with Information and Communication Technologies

"Technology in itself cannot spawn a revolution in educational approaches or results."

(Veenema & Gardner, 1996, p.74)

1.1   How can ICT contribute to the geography curriculum?

There are two relatively distinct ways in which information and communication technologies (ICT) can be incorporated into geography courses:

Teaching ABOUT technology

In this approach, the focus is on introducing students to the technology (hardware, software and data), and what can be done with it. In geography, this has become a common approach during the past two decades, as many departments have recruited 'new blood' in technological areas of the discipline, acquired their own specialist computer resources, and added optional course units in 'geographical computing', 'automated cartography', 'remote sensing', 'geographical information systems', 'spatial science' and the like. There is considerable support for teaching about geographical information technology from a growing number of textbooks, software, off-the-shelf lecture materials (for example, Kemp & Goodchild, 1991), and an ever-increasing number of Web pages (for example, the National Center for Geographical Information and Analysis (NCGIA) Core Curriculum in GIScience [GISCC], and the Remote Sensing Core Curriculum [RSCC]. Recently, a new twist to this approach has been given by Ò Tuathail & McCormack (1998a; b) who suggests the need for students to be trained to use the Internet 'as a classroom', critically examining how its content is shaped by the commercial process of globalisation.

Teaching and learning WITH technology

In this approach, the focus is on developing students' understanding of geography, with computer and communication technologies being used as tools for teaching and/or studying the subject. The educational use of computers has a relatively long history, and is variously known as 'computer assisted learning' (CAL), 'computer assisted instruction' (CAI), 'computer based learning' (CBL) or 'computer supported learning' (CSL). More recently, communication technologies have begun to make major contributions to teaching, first with the development of local area networks within institutions, and more recently with the expansion of the Internet. Despite several national courseware development and dissemination initiatives since the 1970s (for example, the National Development Programme in CAL (NDPCAL), the Computers in Teaching Initiative [CTI] and the

Teaching and Learning Technology Project (TLTP)), the use of technology for teaching and learning has made a much slower penetration into the geography classroom than teaching about geographical technology.

Clearly, there is considerable potential overlap between these two approaches. Computers, for example, can (and are) used to help students learn about geographical information systems (GIS), remote sensing, global positioning system (GPS), and other relevant technologies. However, there seems to be very little movement in the other direction — very little of the expertise in GIS-related technologies at departmental level appears to find its way into the teaching of substantive geography courses. In the rest of this Guide, it is the use of technology by teachers to put across geographical ideas, and by students to help them understand geographical concepts and acquire geographical skills, that will be our main concern.

1.2 What are the most effective uses of ICT in geography teaching?

There can be no absolute answer to this kind of question, because the effectiveness of ICT depends to a large extent on how the technology is used, and this will vary between tutors, between students, between courses, and between geography departments. The best advice from the literature and practitioners appears to be that students should be actively engaged in the learning process, and not treated as passive recipients of information, and that ample opportunity should be provided for discussion. Thus, any ICT resources that facilitate active learning and collaborative study should prove beneficial to student learning.

A broad set of educational roles of the computer in geography is discussed by Unwin (1990), who sees the computer providing the following:

- data and information
- laboratories for investigating the world
- analytical tools
- instruction

At a more detailed level, a range of conventional educational activities can be identified to which ICT can make a valuable contribution:

- course preparation and management
- data exploration
- reading
- expository teaching
- conjectures and refutations
- small group discussions
- problem solving and design
- tutoring and coaching
- writing
- seeking help
- making presentations

Each of these uses is discussed in detail in Gold *et al.* (1991). Unfortunately, neither of the above reviews paid much attention to the educational roles of communication technologies in geography, which is perhaps a salutary reflection on the speed of change in the broader technological landscape impacting on education. We intend to rectify this omission in the rest of this Guide.

1.3   Should ICT be taught as a key skill?

"Technological literacy — computer skills and the ability to use computers and other technology to improve learning, productivity and performance — is a new basic that our students must master."

(Clinton, 1997)

"...the key skills of communication, both oral and written, numeracy, the use of communications and information technology and learning how to learn. We see these as necessary outcomes of all higher education programmes."

(NCIHE, 1997, Introduction, para 38)

As these quotations from President Clinton and the Dearing Report make clear, 'computer literacy' is fast becoming a skill required of everyone going through the education system on both sides of the Atlantic. In the UK, IT skills are now included in almost all official lists of 'basic skills' or 'key skills', typically alongside numeracy and communication skills, in both secondary and higher education. Evidence from recent Teaching Quality Assessors' reports clearly suggests that a growing number of geography departments are emphasising the importance of generic IT skill acquisition by students. Reeve (1985) and Rees (1987) provide early examples of how a 'ladder' of IT skills can be built into a geography curriculum.

One of the reasons why generic skills teaching has become bound up with the use of IT is because of the growing use of versatile 'generic software' such as databases, spreadsheets and word processors in geography courses. Generic software contains no geography at all, and usually no facilities to support teaching or learning. However, it is attractive to geography teachers not only because it can be readily adapted for geographical work (for examples of adapting spreadsheets see Lee & Soper, 1987, and Hardisty *et al.*, 1993; for an adaptation of a statistical package see Griffith, 1992), but also because student familiarity with this kind of software is seen as providing them with considerable advantages in terms of employability.

One of the dangers in focusing too heavily on the provision of ICT skills is that their acquisition becomes the over-riding objective on a course, linked closely to the perceived needs of the job market, and driven by the mainstream products of the computer industry. The role of ICT in supporting the learning process is relegated to secondary importance by its instrumental role in solving real-world (i.e. employment-related) problems. Where this occurs, the ICT component of geography courses tend to become little more than training courses in specific (commercial) computer packages, with far less time being devoted to the underlying principles, applications, assumptions and limitations of ICT. The technology also tends to become marginalised in technology course units, and often fails to break out of its enclave to provide support for the teaching of substantive modules offered by the department.

As a teacher, you should consider your responsibility in this matter carefully, not only to ensure that ICT is used effectively to support the learning of geography, with students acquiring supposedly transferable skills as a by-product, but also so that ICT is introduced in a way that provides students with a critical context for its use in the world beyond academia (see Pickles, 1995). An innovative approach has been taken in the geography department at the College of St Mark and St John in Plymouth, where interactive courseware has been introduced to provide study skills teaching. Here, the computer-based skills teaching is integrated into the geography curriculum across the first two years of the geography degree, where it is linked to substantive geography modules (see Case Study 1).

Case Study 1

Using Courseware to Integrate Skills across the Curriculum

The Department of Geography at the University College of St Mark and St John, Plymouth, has developed a programme of student study skills work as a 'spine' which extends through the first two years of the undergraduate programme. The activities in this spine are introduced through, and run alongside, the substantive geography modules, and involve some additional reflection and organisation on the part of the students. A novel feature is that successful completion of the skills spine leads to a college award in interpersonal and study skills. (Further background details of this course are provided in the companion Guide on 'Transferable Skills and the World of Work' by Chalkley & Harwood, 1998.)

Students are supported by a tightly defined checklist of skills, which incorporates details of the evidence required for the completion of the course. They are also provided with a study calendar, which provides a cross reference between the geography modules (the context for the skills) and the skills checklist. The issue dealt with here is how this skills programme is supported with adequate resources, despite declining staff:student ratios, and other pressures which have reduced staff-student tutorial contact. The resource base is provided by a range of computer assisted learning materials, which have been made available on the college network. These include TLTP software from the *GeographyCal* project, as well as materials designed by college staff. Without these resources, it is unlikely that geography staff could ensure a coherent experience for all students undertaking the programme.

Examples from the programme

The skills programme currently consists of eight study units. Initially, computer assisted study was introduced into just two of these units, in the form of the TLTP module on Social Survey Design. This was used for a two-year period to introduce students to sampling methods and quantitative and qualitative approaches to research. Building on this experience, a computer element has now been embedded into all skills units. Some examples are outlined below.

Essential study skills

This includes use of the library and other sources of information. In addition to the college introduction to the library, students also work through the first part of the TLTP module, 'Making Sense of Information'. In this way, a great deal of basic information is covered by students in their own time, with a minimum of tutor involvement.

Communication skills

This includes the design and selection of appropriate graphical representations of information. As the basis for this, students are required to study part of the early TLTP module on 'Graphical Excellence'.

(cont.)

Information technology skills

In this unit, students must show that they have used the Internet as a source of information. Here, they use the second part of the TLTP module, 'Making Sense of Information', as background preparation for an assignment in which this skill is applied. Students are also required to explain what is meant by a GIS, and to present an example of a GIS used in one particular context. The background here is provided by the TLTP modules 'Introduction to GIS' and 'The House Hunting Game'. (An in-house version of the second of these is used.)

Cartographic skills

Students produce a range of maps (both manual and computer designed) over a couple of months as an element of their first-year geography modules. In order to encourage critical awareness of mapping techniques, they are required to complete the TLTP module, 'Map Design', before putting together a coherent selection of maps for their portfolio.

Strengths and constraints

Students and staff have both been supportive of this new initiative. The institution has validated the programme, and is proposing to evaluate its success with a view to extending the concept to other disciplines. There appear to be a number of reasons for this support:

- The outcomes are clearly attractive to students. They have the opportunity to receive credit and recognition for work accumulated over a period of time, in the form of a certificate and a transcript of their areas of strength. This provides a focus for developing career opportunities and is a concrete source of evidence for employers. While this programme is much broader than IT alone, the extensive use of IT is certainly an important feature.

- The programme is supported by staff who see its potential as a focus for module activities and assignments. Given a reduction in class contact time, some basic concepts and skills can be dealt with outside lectures and practicals.

- It provides tangible evidence of quality provision in the area of interpersonal and study skills.

The programme is not without its problems, however. In particular, there are constraints related to the extensive dependence on IT:

- Staff have to become very familiar with the TLTP materials if they are to reflect their use in modules. This familiarity can take a long time to achieve.

- Many of the materials take students a long time to work through. Unless they have a focused task in mind, student use of the software can become very diffuse and learning may only occur at the surface (if at all). Tasks may be provided, but yet again staff have to become very familiar with the scope of the materials in order to support student learning activities.

- Monitoring and tracking student use of the computer materials is problematic. Do students have to indicate their use of the materials? How should this be shown? How can it be verified? A network monitoring system which allows staff to check on which students have accessed the software (through their unique network passwords), and for how long, might be desirable. But who would monitor this?

- The use of too many software modules (most of which adopt similar styles) can lead to boredom ('TLTP fatigue'). Parts of modules introduced at suitable intervals through the programme may be the answer here. This is something that will be carefully monitored.

- Student access to the materials may prove a problem as computers are in great demand in the college. Will students find time to complete the work required unless some preferential access to the network is provided? Would this be fair on other students?

Innovator: Sue Burkill, Department of Geography, University College of St Mark and St John, Derriford Road, Plymouth, PL6 8BH. Email: stasmb@lib.marjon.ac.uk

1.4 When and where can ICT fit into my existing teaching?

All too often in the past, the advent of a new educational technology has been viewed as a replacement for existing teaching and learning activities. ICT, like other educational resources, has certainly been touted as a teacher substitute. However, it is far more productive and far less threatening if it is seen as offering a variety of contributions to the existing curriculum. ICT may be used:

As a preliminary to other study activities

A common use of courseware is to provide a basic introduction to an area of study (for example, using some of the TLTP modules), which is then discussed and/or expanded on in seminars and tutorials. Computer networks are increasingly being used to disseminate course-related information, such as course handouts and reading lists, for students embarking on conventional forms of study. An increasing number of departments are setting up Web pages that students can browse through to familiarise themselves with a future fieldcourse. Examples of such sites can be found through the CTI Web site [VFCS].

Another useful role of technology is to provide students with training in various skills that they will need on subsequent study activities. For example, the TLTP Soil Surveyor program from the Centre for Land Use and Environmental Sciences simulates some of the situations that students will encounter in the field, and is designed to help them develop basic soil survey skills to a level that will make their actual fieldwork more productive. Warburton & Higgitt (1997) provide a persuasive argument for using IT to help students prepare for geographical study in the field. Drawing on their experience with using GIS and multimedia tutorials in physical geography, they suggest the following preparatory roles for computer-based activities before students go out into the field:

- Introduction to fieldwork themes.

- Provision of regional context.

- Familiarisation with pre-selected study sites.

- Selection of other useful study sites.

- Provision of fieldtrip planning information.

- Briefing on safety and related issues.

- Training in field skills.

As a substitute for conventional teaching

On introductory level modules with large numbers of students, several departments are considering the replacement of lectures by interactive courseware materials, particularly for training in basic skills. With suitable tutor or demonstrator backup, this form of training can be beneficial because it allows students in mixed ability groups to proceed at their own pace, and to undertake as many practice exercises as are necessary for them to grasp the skills being taught.

As an alternative to conventional activities

In appropriate circumstances, students can be offered a choice of how they will study some element of geography. Such a choice enables students to match the available teaching approaches to their personal learning style. In these circumstances, students who feel confident in their ability to learn independently might prefer to use a CAL package rather than attend a parallel lecture. It is interesting that in many evaluations of the impact of ICT on teaching, parallel teaching methods are frequently adopted (for example, ICT and other modes of learning), and yet this approach appears not to be used in routine teaching.

As a component of other teaching activities

ICT can be wheeled into class to provide occasional support for conventional teaching and learning activities. For example, the use of electronic slides, hypermedia, live data visualizations or simulation models in lectures can help staff put across their ideas in a more interesting and/or forceful manner (Conway, 1994). In this context, some research evidence suggests that computer-generated animations are more valuable than static displays, which should give pause for thought to those contemplating replacing overhead foils by computer-generated slides, especially if the latter are simply more colourful versions of the former. Seminars and tutorials, which are often problematic for tutors, can also be given a fillip by introducing the computer or an on-line data resource as an integral element of the discussion. The use of computers in practical classes is now routine, and is discussed further in the companion Guide on 'Practicals and Laboratory Work in Geography' (Birnie & Mason O'Connor, 1998). Gardiner & Unwin (1986) describe the benefits of taking a computer on geographical fieldcourses.

As a follow-up to other study activities

The use of the computer to reinforce learning is at least as powerful as its use to initiate learning. An obvious example is the use of computer facilities to process the results of data gathered in a laboratory or field exercise, though the availability of cheap portable computers now makes it feasible to take such facilities into the field where they become a natural component of other study activities. Another example is the setting of practice tasks for students who have been introduced to a new area of knowledge or skills in a conventional class. With more and more halls of residence being equipped with PCs, and an increasing number of undergraduates having their own PCs at home, computer-based follow-up exercises can now be set for off-campus completion. However, for this to be successful, arrangements will have to be made to provide students with copies of relevant software and datasets. For an alternative application of this approach, in which an entire module of computer-centred work follows a module of lectures focusing on development issues, see Cook (1997).

Case Study 2

Using computer software to provide case studies

A decision was recently taken to use the 'Biogeography' module in the *GeographyCal* software series as a replacement for some lectures in a second-year module on 'Biogeography and Ecology' in the Department of Geography at the University of Leicester. The aim is to teach the broad theoretical perspective in lectures and to use the software, which focuses on heathlands and moorlands, as case studies to illustrate the principles covered in the remaining lectures.

When the lectures were replaced by software, the tutor sat in the computer laboratory to offer students assistance. However, because the modules are relatively self contained, the students' questions were more to do with getting the software to run than on their contents.

This is the first year that this material has been used in this way, and the experiment has thrown up a number of issues which will lead to a modification in the way they are used in the future. Two of the more interesting are that:

- Students do not necessarily turn up for the scheduled sessions for using the software. Several used the material outside the lecture slot which was allocated to them (which is fine), and others ran through as many sessions as they could in one sitting, rather than focusing on the module which had been indicated (this was not considered educationally valuable).

- Students tend to copy down onto paper every screen presented by the software, spending a considerable amount of time copying the material but not necessarily comprehending it.

Nevertheless, in the final exam for the module, most students used case studies on heathlands and moorlands derived from the software as examples. Also, when most of the students attended a fieldtrip in the Peak District, as part of another second-year module on environmental techniques, they were able to interpret the vegetation on the moorland quite successfully, despite having only encountered descriptions on the software module previously. This evidence seems to suggest that although students do not always appear to use the software in the way that tutors might have wished or planned, they do appear to get something useful out of using it.

Innovator: Jane Wellens, Department of Geography, University of Leicester.
Email: jw27@leicester.ac.uk

1.5   Who should use ICT on geography courses?

Two broad strategies for using technology for educational purposes can be defined, based on who is in control during the learning process:

Teacher-centred   Lecturers use computer and communication technologies in various ways: as a lecturing aid, as a tutorial companion, or as a means of disseminating information to students. An example of the first of these uses is the multimedia 'lectureware' developed by Krygier *et al.* (1997). See also Webb (1997).

Student-centred   Students use computers and communication technologies in various ways: as a source of reference information, as a substitute for formal teaching, as a medium for discussing or collaborating with other students, and as a medium for undertaking follow-up exercises after formal classes. An example of the last of these is the series of multimedia 'enrichment' exercises developed by Proctor & Richardson (1997).

A word of caution is necessary here. It is often assumed that by providing students with computer-related work, one is automatically handing over to them the control of their own learning. However, this is not necessarily the case. With many forms of computer assisted learning, there is a danger of replacing teacher control of the learning process by computer control. Tutorial style software, for example, can hold a tight rein on student learning, and many simulation and data analysis programs are designed in such a way that students are unable to exercise more than a minor degree of initiative and freedom in exploring geographical ideas and information.

It was an awareness of the straight-jacketing effect of certain kinds of computer work that led Papert (1980) to pose his famous question: 'Should the computer program the student?' This led to his development of educational 'micro-worlds', which are computer environments that encourage students to devise their own learning strategies. Examples of such resources include models and simulations, interactive hypermedia, and action learning games, all of which can provide environments in which students can work out their own routes to geographical understanding.

But the existence of flexible, student-centred computer resources is not in itself enough to transfer control of the learning process into student hands. Even when computer resources are designed for flexible use, their deployment is still determined largely by teachers. For example, data analysis, mapping, interactive video, expert systems, simulation and modelling software can all be used in a teacher-centred way, and this can be a significant barrier to student learning. This does not mean that tutors should be excluded altogether; they often have a significant role to play, even in student-centred learning activities. McGee & Boyd (1995), for example, describe the positive roles that can (and should) be played by a facilitator of computer-mediated discussions among teacher trainees.

In practice, of course, the distinction between teacher-centred and student-centred uses of technology is often blurred. Even when students take charge of the technology, there is usually a hidden educational agenda imposed by the exercises they are asked to undertake or, less obviously (and perhaps more insidiously), by the educational paradigm embedded in the design of the software they use (to be discussed later). Also, teachers are sometimes unwilling to 'let go' of the technology, even during student-centred uses. An example is the current debate amongst academics over whether undergraduate students should be allowed to place their personally authored course materials on a departmental or institutional Web server. (This issue is discussed further in Section 3.4.)

Case Study 3

Using home-made software to illustrate difficult concepts

Mike Tullett, in the School of Environmental Studies at the University of Ulster, has been using some simple, home-grown computer programs (in BASIC) to teach elements of physical geography for the past eight years. His programs model the geostrophic and gradient winds, wave height, duration and time period. The students can enter any pressure gradient at any latitude and run them until they get bored! Other programs model reflection from plane water, the sun's daily path across the sky at any latitude/longitude and date, and there is also one that calculates humidity.

The programs described

For the wind programs, students enter the isobaric interval in millibars, then the isobaric spacing in kilometres and, if they are dealing with a cyclone or anticyclone, the radius of curvature. If they enter impossible atmospheric variables then the program prompts them to be more sensible. Results are largely numeric, though one program gives a graphical plot of geostrophic wind speed against latitude for the same pressure gradient.

The waves programs, which model both shallow and deep water waves, require students to enter a wind speed in m/s, a fetch in kilometres and the duration of the wind blowing in hours. The output includes the significant wave height in metres, wave length in metres and wave period in seconds. Students are asked to keep two of these constant and vary the third, and to do this for all three variables. They are thus able to see that all three have limits beyond which the waves would not grow. By running the program many times they are able to produce their own graphs, say wind speed versus wave height with constant fetch and duration.

The sun program produces graphical output. Students enter day, month, latitude and distance from nearest time meridian (in degrees W or E), and the program plots the sun's elevation on the vertical axis of a graph against clock time on the horizontal axis. (The 'equation of time' is built into the program.) Students are always fascinated when the program is run for the north pole in summer. Very few have ever realised that the sun just seems to encircle the sky at the same elevation over 24 hours, and their concept of time, based on noon being when the sun is due south and at its highest point in the sky, breaks down near the poles.

Motivations

Although the original idea was to put personal programming skills to some use in teaching difficult topics, there was also the educational problem that putting formulae on the board without the students ever putting them into practice produced limited understanding. In the 'good old days', students would be asked to do a hand calculation, but this was out of the question with the wave programs, because they involved hyperbolic functions.

Another fundamental problem is that the mathematics ability of the introductory groups of students is usually very limited, and their background in physics is even weaker. These weaknesses can usually be traced back to the school syllabus.   *(cont.)*

Evaluation

Students say they have found the programs useful in helping them to understand some very tricky concepts, wind modelling in particular. An assessed exercise has now been based around the use of these programs, in which students have to present their numeric answers and also a short written section with graphs to summarise their findings and demonstrate their understanding of the concepts.

Mike's experience stands in clear contrast to the increasingly common use of Windows-based, multimedia-style courseware, which is costly to produce and does not necessarily produce greater educational benefits than simpler computational programs. The cost of producing the software has been very low, except for the time needed to learn how to program, back in the late 1980s. However, a recent twist in this tale is that the acquisition of new computers now makes it necessary to rewrite the software for the Windows environment.

Innovator: Mike Tullett, University of Ulster.
                 Email: MT.Tullett@ulst.ac.uk

1.6   What can ICT do to foster communication in the learning process?

"Personal contact between teacher and student, and between student and student, gives a vitality, originality and excitement that cannot be provided by machine-based learning, however excellent."

(NCIHE, 1997, para 8.21)

In the continuing debate between the merits of teachers and technology, it is generally agreed that human teachers have the edge in fostering relationships in and around the classroom. Many pundits therefore suggest that the teaching of knowledge and ideas should be assigned to the computer, leaving the development of interpersonal skills to the human instructor. However, there is an immediate problem here, because all learning involves an emotional or affective dimension, and a considerable amount of effective learning derives from the emotion that is generated in interactions between learners, and between learners and teachers. Consequently, a crude separation of content learning from interpersonal skills development might not only undermine the effective learning of geography, but it might also prejudice the development of personal and interpersonal skills.

The 'individualised instruction' approach of earlier CAL is no longer seen as the only or best contribution that ICT can make to student learning. Recently, emphasis has shifted towards the use of ICT to enable 'collaborative learning'. Although students will continue to study alone, and ICT will continue to support such learning activities, emerging technologies make possible a growing number of collaborative learning arrangements, which can be used to support a number of relationships between and among learners and tutors: peer group learning, mentoring, apprenticeships and so on. For an example of the use of distributed learning environments in science education, which includes collaborative student study of the weather, see Pea (1993) and Pea & Gomez (1992); for a general study of collaborative and individualized learning see Johnson & Johnson (1975).

Using internal and external networks, the following types of interpersonal communication can be encouraged, both between students, and between students and tutors:

- *One-to-one*

  for example, tutor providing course advice to a student; one student asking another student for help on an assignment; tutor exchanging ideas on a new course with a tutor in another institution.

- *One-to-many*

  for example, tutors disseminating course-related materials to students; student leader distributing work assignments to individual students in a study group.

- *Many-to-one*

  for example, students submitting course assignments to tutor; students sending data to a team coordinator for collation following a field visit.

- *Many-to-many*

  for example, students and tutor holding a multi-campus, networked seminar; tutors discussing course-related problems through a moderated or unmoderated discussion list.

The critical question underlying these possibilities is whether one needs technological mediation for any of these forms of communication, all of which have been used for centuries in education. Certainly, communication technology should not be used just for the sake of it and at least one commentator (Stoll, 1995) believes that the Internet reduces rather than increases human interaction. Individual teachers will need to decide whether it adds a new dimension to current educational practices, or whether it enables new modes of learning (for example, at a distance) to take place.

1.7 How much do I need to change in order to benefit from ICT?

It is often said that technology is only a tool, and that like all tools, its effectiveness lies not in its inherent qualities, but on how it is used. By simply bolting technology onto existing teaching practices you may not achieve too many benefits. For ICT to deliver its potential, you may need to consider making changes in how you deliver the curriculum. Here are some of the changes that might be considered by individual tutors:

- Restructure your teaching timetable to provide longer study periods; these can be used for intensive project-based work supported by ICT.

- Arrange students into small groups when learning new ICT skills (pairs or 'buddies' work well), so that they can learn from one another.

- Establish clearly defined goals for ICT-supported work, to prevent students spending too much time or going off at a tangent.

- Consider the amount of time you spend in front of the class, and increase the amount spent helping individual students with their ICT-supported activities (this support can be face-to-face or by email).

- Introduce a greater amount of active learning into your students' study diet, adopting the many ICT resources that can be used to encourage active problem solving. Then change your role to one of trouble-shooter.

- Limit the amount of time students are required to sit passively in front of a computer screen, either by choosing courseware that is highly interactive or by demanding that students use the computer as a resource.

- Integrate computer-based work with conventional study activities, in order to avoid ghettoising ICT, and to build on the combined strengths of multiple approaches to study (see Case Study 1, Section 1.3).

- Familiarise yourself with some of the key computer and information technologies currently in use (for example, multimedia and the Web), so that you can evaluate what is available in geography, and so you can provide students with broader contextual guidance on their use of ICT.

- Develop your own learning materials to accompany ICT resources identified for student use. For example, Taylor (1997) produced a worksheet to accompany the 'virtual tour' down the Los Angeles River available on the Web [LAR].

- Consider changing your current assessment practices to allow you to adopt computer-based methods that provide better feedback to students and at the same time reduce your assessment load. See Charman (1997) and Chapman (1997) for examples of the adoption of software to provide regular student assessment.

Heads of geography might also consider making the following changes:

- Restructure departmental resources to release funds for innovative applications of ICT to teaching — but make sure these innovations are (peer) reviewed, and that the lessons learnt are disseminated to all staff.

- Subscribe to the TLTP follow-on courseware development effort at the CTI Centre for Geography, Geology and Meteorology at Leicester.

- Broaden the uses of ICT for administering learning and teaching in the department — for example, course management, module evaluation, budgeting. If these applications are highly visible in everyday educational activities, individual staff are more likely to adopt ICT in their own teaching.

- Undertake a 'blitz' on the adoption of ICT in your department, perhaps by adapting some of the ideas used in the 'IT Term' organised at Oxford Brookes University in the summer of 1996. A Web site has been set up with details of that experiment and an independent evaluation [OITT].

- Move away from what Americans refer to as the 'credit-for-contact' model of reward (i.e. based on class contact hours) that currently dominates staff allocation and accounting practices. Consider alternative reward structures that could be adopted at the departmental level that will encourage staff to experiment with the new technologies, based on (say) the learning hours supported.

- Participate in the university's top-level strategic discussions on the introduction of ICT across the curriculum. The lack of a clearly articulated institutional strategy and/or a poorly coordinated delivery at institutional level can often undermine local initiatives. Current thinking (for example, Laurillard, 1993; NCIHE, 1997) is that top-down strategic planning at the institutional level is an essential pre-requisite for successful implementation of ICT.

- Create a departmental strategy for the adoption of ICT over the next 3-5 years. You do not need to take the technological high road; a low-tech strategy may be more appropriate (see Gibbs, 1997 for a valuable example of institutional strategy setting).

The responsibility for stimulating change must also come from higher levels — institutional, national and international — and also from subject interest groups (NCIHE, 1997). Only by a concerted effort at all these levels will higher education be successfully transformed to capitalise on the benefits that the new technologies have to offer. Geographers must be alert to the pressures for change that are being mediated by ICT (Rich *et al.*, 1997).

Ultimately, of course, your use of ICT will stimulate changes of its own, both in and around the classroom. For a penetrating analysis of the way in which technologies rearrange relationships throughout the educational system, see Postman (1995), who argues that "technologies are ecological". The literature on 'business process reengineering' also has something to say on this issue, because it addresses the question of how far the rewards of new technologies to organisations are contingent upon a radical re-appraisal of established structures and modes of working within those organisations.

1.8   How can ICT fit my teaching style and methods?

The short answer to this particular question is that it should be relatively easy to use most ICT resources for your current educational activities (for example, see the list in Gold *et al.*, 1991). Some ICT resources are designed for particular classroom uses, but they can usually be adapted for use in entirely different ways. For example, a simulation program designed for independent use by students can be used by a tutor in a tutorial or seminar to encourage discussion (Unwin, 1997, describes undergraduate and postgraduate use of the same energy balance model). Similarly, a GIS or data visualization package designed for research or project work can be used as a 'live' visual aid to add spice to lectures. This is invariably more effective that the typically passive electronic slide-show; an useful example is provided by Stark (1986). Things you might do include:

- Adapt an item of courseware to your particular style of teaching. For example, try using one of the TLTP *GeographyCal* modules in a way not intended by the authors.

- Adopt an ICT resource for use in an unfamiliar educational context — for example, collaborative learning, distance learning (learning at work or at home), independent learning.

- Search the Web for examples of mislabelled educational materials in geography. For example, how many 'tutorials' can you find that consist simply of a few paragraphs of text and/or pictures; how many 'lectures' out there are little more than a list of bullet points; and how many 'virtual fieldtrips' are little more than a multimedia version of a regional geography textbook or a travel agent's brochure?

- Consider which educational philosophy best defines your favoured teaching and learning methods. Do you have a coherent educational philosophy? Is it necessary?

This leads us naturally on to the next issue, because there is usually some degree of fit between one's educational philosophy and the teaching methods one uses.

1.9   Does ICT suit students' preferred learning styles?

Research into the way students learn suggests the existence of different kinds of learning style, and even different types of learner. Pask (1976; 1988), for example, distinguished between two groups of learner (the 'serial' and the 'holistic'). Most of the current classifications of learning styles (see Felder, 1996 and Schmeck, 1988 for useful reviews) were stimulated by Kolb's 'Learning Style Inventory' (Kolb, 1984) and by Briggs-Myers & Myers (1980), who developed the widely used 'Myers-Briggs Type Indicator'. In recent years, a small industry has grown up in educational psychology which purports to identify learner styles using various discriminatory criteria, and a number of paper, computer and Web-based questionnaires are available to identify these styles (for example, the Keirsey Temperament Sorter [KTS]).

The implications of these findings for ICT is clear in principle, but perhaps less easy to put into practice. In principle, given that there is a diversity of learning styles and/or learner types, different types of computer and communication facilities could be pressed into service to match these styles and types. One way of doing this might be to test all students in a particular module to identify their preferred learning styles, and then match individual students to teaching methods and learning resources that best relate to their preferences. An example of this approach is provided by a module in Computer Information Systems module at the US Military Academy [CIS], which tests students using Felder's Learning Model and then sorts lesson material accordingly.

An alternative approach is to treat student cohorts as a whole, but to ensure that the entire set of teaching approaches and learning resources used by them exercises each of their many possible learning styles. This approach dispenses with the assessment of individual students, and ensures that students exercise a variety of learning approaches during their study, and not just their preferred style. However, it requires careful instructional design by individual tutors, and/or coordination within entire courses or across an entire department. Which approach do you think is most suited to your own course or department?

1.10   How does ICT relate to models of the learning process?

There are several models that can be used to help you decide how ICT can support the learning process. At least two have been designed with ICT in mind: Laurillard's (1993) conversational model, and the learning framework devised at Heriot Watt University to evaluate the potential role of video-conferencing and other technologies in higher education (Coventry, 1997). However, the more general model of the learning process, Kolb's well-known experiential learning cycle (Kolb, 1984), illustrated in Figure 1, can be used just as effectively to contextualise the use of ICT.

Kolb stressed that experience (or doing) by itself is not enough; experience has to be reflected on and built into modified knowledge structures. (As Kolb himself put it: "Learning is the process whereby knowledge is created through the transformation of experience" (Kolb, 1984).) His model thus integrates reflection into a complete framework of learner activities: concrete experience, reflective observation, abstract conceptualisation, and active experimentation. In Kolb's scheme, the learner can enter the cycle at any stage, but must then complete the activities in sequence. As with the other two models, we can associate various ICT activities with each stage of the cycle. Mellor (1991) describes an application of the model in a soil science project, and identifies the study activities (for example, fieldwork, laboratory analysis, data interpretation, writing a report) undertaken at each stage of the cycle.

Figure 1: *Kolb's Experiential Learning Cycle (Kolb, 1984)*

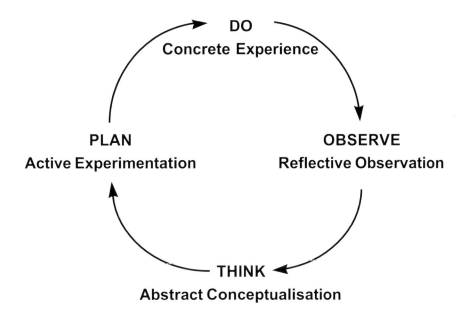

Each of these models provides its own particular insights, and all can be used in some way to guide your use of ICT. However, the over-riding benefits of Kolb's model are that a) it is already well-established and widely adopted, in geography as well as other disciplines; and b) it embraces all learning activities, including those that do not involve the use of ICT. There is probably little in the other two models that could not readily be included in the experiential learning cycle. But, as with all such discussions, you will want to decide for yourself which is most effective for your purposes.

Case Study 4

Developing multimedia courseware to reflect a philosophy of learning

An example of the innovative educational use of multimedia is provided by Veenema & Gardner (1996), who have developed a CD-ROM for students of the American Civil War. The authors argue that students must be challenged to understand multiple meanings of events, and that multimedia can be a powerful medium for doing this. Building on Gardner's earlier work on the concept of 'multiple intelligences', they suggest that since "the mind is neither singular nor revealed in a single language of representation, our use of technologies should reflect that understanding."

Building on this philosophy, they approached the design of the CD-ROM with the principle that students should be introduced to multiple perspectives of a famous battle. They achieved this by providing information recorded by a number of different contemporary observers of the battle, and by providing four 'guided paths' to help students navigate this material. Among the materials included on the CD-ROM relevant to geographers is information on the role played by terrain and weather.

Innovators: Shirley Veenema and Howard Gardner, Harvard University.

Reference: Veenema & Gardner (1996).

1.11   How can ICT foster independent study and student-based learning?

It is difficult to provide unambiguous definitions of 'student-based learning' and 'independent study'. However, we can probably agree that in essence they refer to a shifting of the responsibility of learning from the tutor to the student, and involve students in both setting some of the goals of their learning and managing some or all of the learning process itself. In other words, the key idea is not whether students are independent or their studies are self-based, but rather the degree to which students are encouraged to become autonomous learners. An essential corollary is that teachers must take responsibility for facilitating this process.

One of the dangers in linking ICT to the goal of independent learning is that it can lead to the adoption of a weak form of 'individualised learning'. Proponents of CAL have always claimed that one of the key educational benefits of the computer was its ability to individualise instruction, by providing students with tutorial software that takes them flexibly, and at their own pace, through a body of knowledge. Unfortunately, most of the early CAL software involved a crude form of tutoring, based on the mechanistic technology of programmed instruction (even modern multimedia tutorial software is too often a thinly disguised form of this earlier approach).

In order to clarify our thoughts, we can perhaps suggest that students contribute to, or participate in, the learning process in three ways: as receptors of knowledge, as explorers of existing knowledge, or as creators of their own knowledge. As we move from the first type of learning to the second, the student's role changes from passive to active learning; as we move from the second to the third, it changes from the discovery of existing knowledge structures to the creation of new ones (new to the student, that is), either through the synthesis of existing knowledge or by engagement in problem solving activities. These three types of learning are all found in conventional study activities, but they can also be found in computer-based work, as suggested in the following table:

Passive   Receiving and (supposedly) absorbing knowledge (for example, watching a computer screen during a lecture illustrated with computer graphics or a simulation model).

Exploratory   Exploring information through guided discovery (for example, experimenting with a simulation, exploring empirical data, or participating in a computer-based game).

Creative   Applying knowledge (for example, building a computer model, compiling a word processed essay, designing or optimising some part of the real world, putting together a multi-media presentation using hypermedia resources).

Recent studies suggest that the most effective way you can support autonomous learning is by providing your students with the following kinds of resource:

- unrestrained access to relevant information

- varied and flexible learning materials (including geographical databases, analytical software and so on), which encourage the creation and testing of geographical hypotheses, and the acquisition of knowledge through a process of conjecture and refutation.

- flexible modelling environments (what Seymour Papert called 'microworlds'), to enable students to explore ideas and alternative scenarios through a process of 'what if?' experimentation.

- communications facilities, to enable students to acquire information, feedback, advice and support from a variety of co-leaner, including tutors, other students, and experts elsewhere.

As we will see in Section 3, this kind of support can be provided by (some) multimedia courseware, but increasingly by access to the Internet and the World Wide Web.

1.12   Does the use of ICT encourage 'deep learning'?

The 'surface' approach to learning is typically associated with students memorising facts, and focusing on passing examinations. The 'deep' approach to learning, by contrast, is usually internally motivated, and is based on students wanting to understand and wanting to find meaning in what they are studying (Entwistle & Ramsden, 1983).

As with other study methods, the learning outcomes from using ICT often depend on the way one uses the technology, rather than on the technology itself. An instructive comparison can be made with fieldwork. At one extreme (for example, the traditional Cook's Tour), fieldwork can develop superficial learning and foster boredom with both the subject matter and the learning process. At the other extreme (for example, where fieldwork is cast in a problem-oriented framework), field study can stimulate student engagement with the subject matter, and encourage deeper understanding (see Bradbeer, 1996). This suggests that we should perhaps not think in terms of 'fieldwork' as a study method in its own right, nor should we think of ICT as a learning method. It is how we package the student's total learning experience that will determine whether we deliver or fail to deliver deep understanding.

In general, deep learning can be encouraged by stimulating students to shift repeatedly between experience and formal learning. A key element in this framework is that students begin with experience rather than with information or theory, and that they continually think about what they are doing ('thinking in action') and also on what they have done ('thinking on action'). The key element of this is reflection, which involves "internally examining an issue of concern triggered by an experience, which creates and clarifies meaning in terms of self and results in a changed conceptual perspective" (Boyd & Fales, 1983). Clearly, using ICT in various ways and contexts can help to ensure that reflection takes place but, as Monke (1996) suggests, putting students in front of a computer can often result in quiet contemplation being squeezed out by task-related activities that have a low learning potential.

1.13   What should I look for in 'good' educational software?

There is no hard and fast answer to this question, partly because there are so many different kinds of educational software and related computer resources available, partly because such materials can be used in so many different ways, and partly because 'good' is such a subjective notion. There are a number of books that provide detailed checklists of the criteria you might use to evaluate available courseware, and there is a valuable discussion of this issue in the report from Heriot Watt University (Stoner, 1997).

A clear distinction should be drawn between two groups of software. The first kind have an instructional element built in, and are designed explicitly to help students learn some element of geography; several of the TLTP geography software modules take this form. The second kind consist of 'generic' software (for example, databases, spreadsheets, word processors, and expert system shells), which are use-free, and other information resources that are relatively independent of particular software (for example, information on the Internet and the Web, or a CD-ROM atlas or multimedia encyclopaedia). Focusing on the first of these (loosely referred to as 'tutorial software'), we can outline some of the key characteristics you should look out for.

- *Interactivity*

  Does the software encourage the student to learn through active interaction with the material presented, or does it simply present pages of information on screen, reducing the student to a passive recipient whose only activity is page-turning? Critics of the passiveness involved in TV as a medium (for example, Postman, 1985; Gilder, 1992) suggest that the new interactive media will promote more effective learning. In practice, however, most interactive software, whether driven by commands or point-and-click interfaces, engage students as reactive rather than initiating partners in a learning enterprise. Heath (1995) argues that so-called 'interactive multimedia' software incorporates only a very weak form of interaction, and fails to provide the rich form of interaction which characterises human communication. She suggests that a more effective alternative could be provided by multicast multimedia conferencing over broadband networks, accompanied by a shared electronic whiteboard, which would facilitate teacher-student interaction at a distance.

- *Flexibility*

  Does the software permit students to follow a variety of routes through the material on offer; does it allow students to switch from a 'beginner' to an 'expert' mode; and does it allow them to skip elements they may already have learnt? It is still too soon to expect tutorial software to exhibit even a small fraction of the infinite flexibility of the human teacher, but modern courseware should avoid straight-jacketing students and be able to go some way towards accommodating their personal learning styles. At the other extreme, of course, too much flexibility can be counter-productive, as we will see when we come to discuss multimedia and the World Wide Web. Some degree of structure and constraint is clearly important, especially to students who are new to a subject or new to the tutorial style of courseware.

- *Feedback*

  Does the software provide students with regular feedback on their study activities? Such feedback is most normally provided in the form of immediate judgements or comments on student's answers to specific questions, but it can also be provided as end-of-section or end-of-module summary evaluations. The style of feedback provided is often tied to the activities undertaken by the student or the questions posed by the software. If an exercise is rather mechanical (for example, learning factual information) then the software may provide intellectually unstimulating questions of the Yes/No variety. By contrast, if the software provides a simulation model that the user has to calibrate to achieve a particular outcome, or it presents some data that the student has to visualize in some way, then the feedback can be more varied and stimulating. Better courseware links feedback in one section of the learning activity to subsequent exercises, and/or uses student responses to guide the student to subsequent activities.

Recently, with the spread of computer networks, a fourth ingredient is being added to educational software:

- *Communication*

  Does the software enable students to share ideas with other students during computer-mediated study activities, and does it permit students to ask tutors for clarification or guidance while they are on-line? Some developers are coming to the conclusion that the ability to communicate with other humans while using educational software is perhaps the most important ingredient of all, and this is something we will discuss further in connection with the Internet and the World

Wide Web. As yet, however, few examples of this kind of software have been developed. A valuable role model could be provided by some of the networked games that are being produced by commercial companies.

Of course, the existence of these characteristics does not guarantee the educational success of a particular item of educational software; effectiveness is often determined more by the use you make of the software. Even the best educational product can be ruined by poor use, just as a poorly designed program can bring benefit to students if it is adopted in an inspired way. It is also worth noting that the educational model or paradigm incorporated in the courseware can also have a significant effect on its suitability for particular students and on its effectiveness on a particular course, as we saw in a previous section.

1.14 How can I confirm that using ICT is benefiting my students?

There has been a good deal of excitement amongst educational technology developers and those with their hands on educational purse-strings about the potential of ICT to improve the learning experience of students and to deliver courses more economically and flexibly than traditional methods. In reality, however, the benefits of using ICT for teaching are hotly debated. On the one side are the developers and boosters who see the benefits as almost self evident, and consequently make exaggerated claims for technology; on the other side are the critics (for example, Oppenheimer, 1997) and practitioners who see little hard evidence to support the assumed benefits.

To be sure, there are a few meta-analyses (for example, Kulik & Kulik, 1987; 1991) which indicate positive learning gains from student use of ICT. But the results of such analyses are hotly contested, and there are all too many evaluations of computer-based innovations in education that fail to prove that the technology improves student learning performance in a measurable way compared with traditional teaching methods. There are also persuasive historical reviews (for example, Cuban, 1986) which place modern ICT at the end of a long line of other technological innovations in education. These suggest that ICT is part of a recurrent educational marketing cycle (represented in the current century by books, radio, television, and cable), which succeeds in part because each generation fails to recall the limited gains provided by the previous generation's innovations. (Russell, 1995 has compiled a devastating catalogue of studies dating back to the 1920s which indicate little support for the assumed benefits of a succession of educational technologies.)

If you are one of those teachers (hopefully a majority) who like to be certain that their teaching innovations are worthwhile, you will want to evaluate your use of ICT in some way. We offer below a few general guidelines drawn from the literature; a more detailed review of methods is provided in the chapter on evaluation in Gold *et al.* (1991).

- Formal approaches to evaluating the educational use of ICT involve comparing a traditional approach to teaching some element of geography with a technology-based approach, or comparing student performance before and after the use of the technology-based approach. In both cases, it is usual practice to divide students into experimental and control groups, and to use a quasi-experimental methodology to compare the two sets of performances. However, it is unlikely that such evaluations will result in a clear 'winner'; the published results of such contests in the literature frequently reveal rather inconclusive evidence for the superiority of ICT over other teaching and learning methods. An evaluation of multimedia materials developed for geography teaching (Proctor & Richardson, 1997) suggests that they did not deliver the expected learning outcomes.

- Don't rely merely on collecting 'hard' data about your use of ICT, and subjecting them to statistical analysis. Most teaching innovations can be effectively evaluated using qualitative data and 'soft' or informal methods of analysis. Some effective evaluations of CAL have been based on gathering protocol information — for example, compiling structured transcripts of students' talk-aloud views of their learning experiences. These often provide a rich source of evidence for identifying the strengths and weaknesses of the particular study approach being used, and for improving subsequent practice. In the Proctor & Richardson study (1997), qualitative evidence revealed that students responded positively to the use of multimedia courseware, even though the statistical evidence indicated little in the way of clear-cut learning benefits.

- When you evaluate ICT, use methods that are sensitive to the broader educational goals and context of your course and/or department. It is often a mistake to take the classic approach to experimental evaluation, and try to isolate what goes on in a particular classroom session or an individual module. Student learning is highly complex, and occurs across many modules and study contexts. It may therefore be futile trying to separate out a particular use of ICT in one of your geography classes from these many other influences.

- Be alert to the possibility of unintended, but positive benefits of your use of ICT. Even when your use of the technology does not seem to have paid off in the ways you anticipated, this does not necessarily mean that the experiment was entirely worthless. Many evaluations of technology-supported learning reported in the mainstream CAL literature suggest that the unintended consequences of particular ICT activities — such as increased student motivation for study — can sometimes be more significant than the outcomes that were intended when the activity was initially planned.

- Be realistic about the time needed to evaluate a ICT innovation properly. Indeed, it may be worth continuing with your use of ICT in the absence of hard supporting evidence, based on your gut feeling that it is rewarding to students. You can be confident in the knowledge that most evaluations tend to show that educational uses of ICT do not normally lead to poorer student performance.

- Don't just single out ICT for rigorous examination — cast your critical eye equally over the other teaching methods you use. You may find that it is not ICT that needs to be justified, but some other teaching method that you have been using uncritically for several years.

1.15   What costs are involved in adopting ICT?

"The adoption of Communications and Information Technology (C&IT) is too big, too expensive, and too fundamental to the operation of the institution as a whole to be decided at faculty level."

(NCIHE, 1997, para 13.22)

The cost of producing high quality courseware is very high. Even in the business world, developing multimedia training materials can prove prohibitively expensive for all but the largest corporations. This means that individual geography teachers or departments are unlikely to be able to justify the cost of courseware production in-house, and should therefore consider acquiring available materials from external sources or joining a courseware development group (such as the TLTP *GeographyCal* consortium) to produce new resources. However, this may mean that you have to accept a less than perfect fit between the specific needs of your own courses and the capability of more broadly-based courseware. Adoption of nationally produced courseware may also involve succumbing to an implicit common curriculum (see Bradshaw, 1995).

Another important dimension of cost is that the purchase of computer-based educational materials almost certainly involves a trade-off against other learning resources. Geography departments and individual tutors therefore need to make a clear-headed choice between alternative learning resources on the grounds of their educational effectiveness, and not simply on the basis of their comparative costs. However, undertaking such evaluations will take time and experience, and may mean that your department provides resources and/or time allowances to enable staff to evaluate educational resources as they appear. Although you may be able to benefit from the experience of colleagues in other institutions, and from reviews in the literature, it will rarely be appropriate to rely solely on evaluations undertaken by others, if only because of the diversity of geography courses in higher education.

In general, you should resist the temptation of adopting ICT simply to save staff time. Although the use of computer facilities can deliver clear educational benefits, it usually requires an investment of time and effort, at least over the short term, to deliver these benefits. For example, staff may need time to acquire the skills necessary to develop computer resources in-house (for example, in writing objective tests or HTML documents). They will certainly need time to evaluate, test and implement bought-in computer resources, and additional time to become proficient in their use.

1.16   How can I avoid failure?

Experience suggests that attempts at using ICT for teaching and learning might fail, or at least not live up to expectations. There are several reasons why this might happen. First, things may go wrong when you or your students attempt to use computer or communications kit for study activities (Flowerdew & Lovett (1992) provide a down-to-earth review of things that can go wrong in the classroom, and suggest helpful remedies). It is generally agreed that thorough preparation and local technical support can help prevent most disasters, but you should draw up contingency plans (for example, creating paper handouts or overhead foils as backup for computer-based materials) in case (or when) things go wrong.

A second reason for failure is lack of suitable support for your ICT experiments: inadequate funds to acquire suitable hardware and software; no teaching remission for the evaluation, testing and/or development of learning materials; and limited access to technical support staff. In relation to the last of these, Browne (1992) has drawn up a summary of the relative merits of using departmental and faculty-based technician support for a course in GIS.

Over and above the practical problems confronted in the classroom are a number of more fundamental reasons for the possible failure of ICT in an educational context. For example, ICT is likely to fail when:

- No attempt is made to change the existing curriculum (for example, course content, teaching styles, assessment methods and procedures) when ICT is introduced.

- Teachers are poorly prepared for computer-based classes.

- Students are poorly briefed and/or de-briefed in relation to computer-based activities.

- The computer is expected to contribute in ways that are outside its competence (for example, as an intelligent tutor).

- The computer is brought in simply to motivate students.

- The computer is adopted primarily to reduce or avoid class contact.

- Departments dabble in courseware development projects for reasons of prestige, politics or funding, or teachers dabble in technology because it gives them a buzz.

- Students are forced to use computers and communication technologies in special rooms on campus, rather than being provided with such facilities wherever and whenever they wish to study.

- Computer-based teaching is kept separate from other teaching and learning methods.

- The use of computers in geography is 'ghettoised' in specialist courses in GIS, remote sensing, and the like.

- Expectations are unrealistically high.

1.17   How should I get started?

If you have not yet grasped the nettle of ICT, but are interested in taking the plunge, then the best advice is to start with something small and relatively risk-free, and then to move on to something more adventurous or more extensive when your first experiment succeeds. Here are some possible starting points:

- Use your World Wide Web browser (for example, Netscape or Internet Explorer) to explore the many geography resources listed in the CTI Centre for Geography's Web site at Leicester University [CTIG]. (If you do not know how to access the

Web, drop into a colleague's office along the corridor and ask them to do it with you.) Adopt some of the material found on the Web for use in one of your classes, either by converting it into a handout, or by giving students the Web address so they can read it for themselves.

- Send an email message to a colleague who teaches similar aspects of geography in another institution. You could ask for advice in solving a course-related problem, or simply share your experience of teaching a particular course. If you do not know the email addresses of colleagues teaching elsewhere, you can find most of them by visiting the Web sites of individual geography departments. (The Web addresses of UK and overseas geography departments are listed on the CTI Web site mentioned above, in its 'Information Gateway' section.)

- Get students on an introductory course which you teach to study a particular topic using the geographical courseware provided by the TLTP Project. (These can be set up on departmental PCs, or on a local computer network.) Rather than add this study activity to students' existing work on the course, consider replacing one of your weekly lectures with the computer-based work. But be sure to set aside part of a seminar or tutorial following the assignment to assess and extend the students' understanding of the material.

- Require students on one of your smaller optional courses to present an assignment in class using illustrative materials (for example, images, animations, text) delivered on a PC. To do this you will need a single PC in the classroom (a portable would be ideal), and a means of ensuring that the screen material is visible to all students (for example, multiple TV monitors, or a 'PC viewer' attached to an OHP). A single, standard item of software should be used by all students (for example, a presentation package or a Web browser), and students should be required to load their material onto the PC at least 24 hours before the presentations are scheduled. (The assistance of a departmental technician can ensure that things go smoothly from a technical point of view.)

If you have already travelled some way down this road, you may want to try out some of the other uses of ICT discussed elsewhere in this Guide.

Case Study 5

A Computer-supported Independent Study Module

At Middlesex University, a semester-long module in Environmental Monitoring Technologies was introduced in the second semester of the 1994-95 academic year. Because there were too many existing modules, it had to be added to the department's course portfolio as an independent study module. There would be no formal teaching; instead, computer-based work would be a major part of the students' learning experience.

Because of the considerable time needed to create effective tutorial courseware, it was decided not to deliver the module content by computer (for example, by developing interactive tutorial software). Instead, it was decided to present information to students using conventional printed materials, and to help them understand and absorb that material by requiring them to undertake practical exercises on the computer. In other words, the computer would be used for what it is usually best at doing: providing a laboratory for student practical work involving the active exploration of information.

Some two dozen 'study units' were created, each dealing with a clearly-defined topic, each including some essential reading and practical exercises, and each requiring between one and four hours of student time to complete. The laboratory element includes a considerable amount of computer-based image processing, and involves students downloading satellite data using a roof-mounted receiver. The assessment is entirely by coursework, and includes a class presentation and two written reports.

The main learning strategy adopted to encourage active learning was the inclusion of a number of questions in each study unit, and empty 'boxes' in which students were required to provide their observations, interpretations and answers. In order to ensure that they did not skim through the material, students were issued with one study unit at a time by the laboratory technician, and could only get the next unit when all boxes had been satisfactorily completed.

The following lessons can be drawn from running this module over the past five semesters:

- Independent study, supported by computer work, can deliver a number of benefits for students. The main ones reported by our students are flexibility of study (in particular, the ability to fit work on this particular module around other course commitments), and the ability to work at one's own pace. Other advantages reported include: variety in patterns of study across a student's entire course; experience in time management and self motivation; and (for some) a greater enjoyment of the learning process.

- Computers do not have to be used in a tutorial mode in order to provide valuable support for independent study. Indeed, their use for data interpretation and analysis exercises can often provide greater scope for active learning by students.

- A module of this kind, which is based on specialist image processing and GIS facilities, and which requires students to make regular use of computers throughout the semester, could not be run without adequate technician support. Preferably, the technician should have some experience of the subject matter being studied.

- Students can easily feel 'cast adrift' when taking an independent study module, even where there is a supportive laboratory technician on hand. So it is important for the tutor to keep in regular touch with them. However, this requires multiple visits to the computer laboratory because, due to the independent-study nature of the module, there will rarely be more than two or three students working in the laboratory at any one time. Email facilities can also be used to reduce the sense of isolation, but face-to-face contact is also very important.

- A considerable amount of thought must be devoted to providing formative feedback to students as they work through resource materials on an independent study module of this kind. Where computer software is not used to deliver information and check

(cont.)

student responses (as in this case), it is essential that some other method be used, to prevent students feeling that they do not know whether they are on the right track. In our case, we were fortunate to have a technician who was expert in the study matter of the module. Elsewhere, module tutors may have to arrange regular meetings with students to go through their completed exercise material with them. A third alternative is for staff to provide individual feedback using email facilities. The underlying problem for us was that the module was implemented on the basis of a minimal allocation of tutor hours, so there was little scope on the work programme for providing regular student feedback. This issue is still the most problematic for independent study modules, particularly where (as in our case) they are introduced as an 'efficient' form of course delivery.

- Students will often work together on their computer-based practical work, despite the label 'independent study'. In general, study collaboration is encouraged, both as a way of mitigating the loneliness factor, and also for improving the learning process. However, it is discouraged during the preparation of assessed work (although this is not very easy to enforce in practice) because our assessment tests individual student understanding of the course material. Although some form of group assessment might be attractive in these circumstances, it can be difficult to implement in an independent study module where students have limited contact with one another. Again, email might be a solution to this particular problem.

- Independent study can expand the choice of modules available to students, and they can usually be offered in both semesters of the teaching year, without a significant increase in teaching hours for staff. However, the up-front cost of developing such modules must be underwritten by the department, the institution, or by a nationally funded project, and the considerable amount of staff time needed to get such modules up and running must be recognised in some way. More importantly, the 'savings' in staff time must not be completely allocated to other activities. In particular, some of the hours freed up by transferring learning to computer-based materials should be allocated to staff so they can communicate regularly with their students during the learning process. The worst form of student independence, which must be avoided at all costs, is that form which is a flimsy camouflage for student aloneness.

Innovator: Ifan D H Shepherd, Middlesex University.
Email: I.Shepherd@mdx.ac.uk

Reference: Shepherd, I.D.H. (1996) The Self-Study Approach to Course Delivery: developing a computer-supported Geography module, *GeoCal*, 14, pp.6-9.

2 Making the most of Information and Communication Technologies

"What's wrong with education cannot be fixed with technology."

(Steve Jobs, co-founder of Apple Computers, 1996)

This section of the Guide will focus on answering three related questions: how can ICT help you and your students to...

- improve existing practice?

  (for example, using computer demonstrations within a lecture)

- solve specific teaching and learning problems?

  (for example, providing remedial tuition for large numbers of students)

- approach the study of geography in new and beneficial ways?

  (for example, enabling students to undertake inter-institutional collaborative projects)

Some two dozen potential contributions of ICT to the teaching and learning of geography are described in this section which attempt to answer one or more of these questions. (The contributions are listed in full in the Contents section.) Each suggestion represents a particular problem that can be solved, or an opportunity that is made possible, by using ICT. Our aim in presenting these is to prove Steve Jobs wrong — or at least not entirely right! For each contribution, two summary sections are provided:

- A contextual statement — the problem or opportunity that may be addressed by ICT;

- A list of possible applications of ICT that can help to solve the problem or realize the opportunity.

Cross-references are provided to related discussions in other Guides, and pointers are included to case studies included elsewhere in this Guide. References are given to relevant discussions in the literature and to additional information provided in the Resources Database. It should be noted that the contributions are not arranged in any order of precedence, and it is to be expected that not all contributions will be equally valuable to all readers.

The focus of each of these contributions is on a teaching or learning issue, and various technological (and conventional) solutions are identified which address each issue. If you are interested to know how a particular kind of technology can be used for educational purposes, then the following cross-reference index may prove useful.

Technology	Contributions
Presentation software	*1*
Word processing	*24*
Electronic essays	*14*
Computer mapping	*1, 5*
Computer simulation and modelling	*1, 2, 4, 5, 8, 9, 13, 16*
Tutorial software	*1, 8, 11, 15, 16, 17*
Skills training software	*2*
Digital datasets/data analysis	*6, 9, 16*
GIS	*4, 15*
Computer games	*4*
Electronic journals	*19*
Bibliographic databases	*19*
Computer-marked tests and evaluations	*24, 25*
Digital videos	*19*
Computer managed learning	*16, 23*
Portable PCs	*10, 13*
Data loggers	*10*
Email communication	*1, 11, 12, 20*
Course materials on local network	*1, 3, 19, 22*
Internet datasets	*3, 6, 7, 21*
Internet tutorials	*21*
Internet collaboration	*3, 11, 12, 13, 21*
Internet visits/fieldtrips	*2, 3, 7, 13*
Internet presentations	*14*
Live satellite data/Web cams	*7*
Remote computer access	*11, 18, 20*
Satellite communications	*13*
Distance learning	*18*

Contribution 1:   Enhancing teaching with technology

Context

Lecturing is still the mainstay of teaching in most geography departments. In the face of declining resources and increasing student numbers, this form of teaching is viewed as a cost-effective means of providing basic orientation in most aspects of geography. However, as is more fully explored in the companion Guide on 'Lectures in Geography' (Agnew & Elton, 1998), lecturing does not always produce the same kind or level of learning as can be achieved by other teaching methods. Here, we will consider how ICT can be used to improve the art of lecturing.

Applications

- At the simplest level, conventional lectures can often be improved by using computer software to produce better versions of hand-drawn overhead slides. Many geography staff are exploring the use of commercial presentation software to replace slides and overhead foils in class, sometimes following experience gained by using the technology for conference presentations. If you have not already started down this route, you might consider converting your lecture materials to Web (i.e. HTML) format, and use a Web browser to display them during the lecture. You will be in good company: software developers like Microsoft are currently switching their help files from original proprietary formats to HTML.

- Lecturing to ever-larger groups of students can be improved by using computer-delivered problems or case studies. These may be worked through 'live' by the tutor, or replayed from recorded sessions. Stark (1986) describes a 'star' lecturer in the US who maps historical census materials in lectures, both to stimulate student interest in the subject matter and to elicit their active participation in class. If you already break up your lecture slots (see Gibbs & Jenkins, 1984), then it will be relatively easy to introduce computer interludes of this kind.

- Help students to learn at their own pace by placing lecture notes on the Web or on a local network. This way, students can re-visit the material at their own pace after the lecture, either to clear up misunderstandings or to fill in gaps in their notes.

- Replace part of, or entire, lectures with active learning exercises (for example, using simulation or tutorial software) to improve student understanding of geographical concepts or ideas (see Contribution 7). Evidence from a number of investigations suggests that carefully designed computer software can help students understand issues that would otherwise remain clouded, particularly if they are able to experiment actively with the materials in their own time. See the example provided by Jane Wellens in Case Study 2 (Section 1.4), and the Web-based course materials created by Iain White at the University of Portsmouth [UPCM].

- Use tutorial and simulation programs in follow-up sessions to reinforce material introduced in lectures. However, care may be needed to avoid making these exercises too mechanical and therefore supportive only of 'surface learning'. Bear in mind that the evidence from controlled experiments does not unequivocally show that students recall information from computer-delivered materials better than they do from conventionally studied materials.

- Lectures may be improved by using materials from other sources. An increasing number of 'lectures' are available on the Web, though *caveat emptor* is the appropriate advice, because a 'lecture' on the Web can mean anything from a sketchy course outline or a *PowerPoint* slide of bullet points to a full-blown set of teaching resources. Examples include the Earth Sciences course on Earth Catastrophes at the University of California at Santa Cruz [UCES], the well-known NCGIA Core Curriculum in GIScience [GISCC], and various geography materials listed in the World Lecture Hall [WLHG].

- Incorporate guest 'lectures' using email (Bailey & Cotlar, 1994), and invite students to discuss the ideas with the lecturer, again using email. This idea can be extended to electronic forums and panels by using a shared mailing list moderated by a lecturer.

One of the main implications of adopting these approaches is the need for thorough-going training, both for tutors and students, in the use of appropriate computer facilities.

Contribution 2:   Reducing costs

Context

The declining unit of resource in higher education is forcing many geography departments to reduce the overall cost of delivering the curriculum. Cost savings are being examined in relation both to staff wages and to general expenses. Among the high-cost study activities in geography are:

- Fieldwork, especially residential fieldcourses and overseas field visits (travel and accommodation costs)

- Laboratory work (technical support and consumables)

- Small-group teaching (staff time)

- Remedial teaching (staff time)

- Marking coursework and exams (staff time)

- Print-based readers and study guides (reproduction costs)

Applications

Ideas for reducing costs in each of these activities are described in several of the companion Guides. ICT can help reduce costs in several ways:

- Supporting the teaching of large numbers of students in effective mass lectures (as described in Contribution 1).

- Reducing the cost of laboratory-based study activities by using computer simulations to replace the use of expensive consumables in laboratory experiments. This has been done in other disciplines. For example, microscopes and slides are replaced by the Open University's Virtual Microscope [OUVM], seismic kit is dispensed with in the Virtual Earthquake software produced by the Electronic

Desktop Project at the California State University, Los Angeles [EDP], astronomical equipment is substituted by software in NASA's SkyView Virtual Observatory [NSV], and animals are replaced by the Virtual Frog Dissection Kit from the Lawrence Berkeley National Laboratory [VFDK] and the Interactive Frog Dissection Tutorial at the University of Virginia [IFDT].

- Reducing the cost of fieldwork by replacing some of the time spent briefing students in the field by computer-based preparatory activities. For example, John Grattan (jpg@aber.ac.uk) at the University of Wales, Aberystwyth, has developed a Virtual Fieldwork site by placing briefing documents on a college intranet to brief students for their field visit to Malta [VFMA]. Using such a resource, tutors involved in team field teaching can also be brought up to speed on a new fieldcourse venue without the expense of a prior visit. Another approach is to use computer software to train students in necessary fieldwork skills (for example, by using the Soil Surveyor software described elsewhere).

- A more ambitious approach is to replace fieldwork altogether by using surrogate field studies. An example from the neighbouring discipline of archaeology is provided by the *WinDig* archaeological dig software. Given that some geography departments spend as much as half their annual educational budget on fieldwork activities, it is hardly surprising that some are looking seriously at 'virtual' fieldwork. On a Web site at the University of Texas at Austin, for example, there are the beginnings of a project to allow students some of the experience of visiting a field site without having physically to travel there [VGD]. There is also a group development project currently under way in the UK to develop a 'virtual fieldcourse' for geography [VFC], and there is a growing number of virtual field trips on the Web in cognate disciplines such as geology [GFT]. Such facilities are seen as enhancing and extending conventional fieldcourse activities, and should not necessarily be seen as a replacement for the real thing. (See the companion Guide on 'Fieldwork and Dissertations in Geography' by Livingstone *et al.*, 1998.)

However, ICT has its own set of associated costs, not all of which are represented by the capital costs of acquiring hardware and software (see the discussion in Section 1.15). Using ICT to reduce existing costs should therefore be considered only when:

- A special grant or fund can be used to implement the required ICT solution.

- Some element of the ICT alternative can be off-loaded onto central budgets. Admittedly, this may not lead to 'real' savings within your institution, particularly if departmental budgets are top-sliced to provide central resources, nor is it likely to be a major tactic in an era of devolved budgets. Nevertheless, it is an approach that has served some geography departments well over the years.

- Effective software and/or data can be acquired for minimal cost from external sources (for example, the Internet, or a TLTP project).

- Local technical support is already available for student and tutor use of ICT.

Contribution 3:   Motivating students

Context

Some topics in geography are perennially viewed as 'difficult' or 'boring' by students. These include technical skills (such as sampling, or statistics), and certain theoretical and conceptual ideas (such as post-modernism, or various aspects of the methodology and philosophy of geography). It is frequently suggested in the mainstream educational literature that computer-based activities can motivate students to learn, but such claims are rarely supported by hard evidence. One of the key studies undertaken in this field (Malone, 1981) cautioned against using 'extrinsic' motivators in educational software; motivation should be intrinsic to the learning activities, or else any motivational effects will rapidly wear off. Perhaps the main message here is not to adopt ICT support for study activities primarily for their motivational impact.

Applications

A number of ICT-based activities have been reported as contributing to student enthusiasm for their studies:

- Introduce material gleaned form the Internet to increase the topicality of geographical courses (see Contribution 6).

- Use simulation, data analysis or mapping software to make tutorials and seminars more interactive (see Contribution 4).

- Use the Internet to visit other parts of the world (for example, reading photographic and video 'essays' and travelogues written by local experts). A number of commercial organisations are setting up 'virtual expeditions' to different parts of the world (for example, TerraQuest [TQ]).

- Use the Internet to enable students to collaborate with students elsewhere in the world (see Contribution 13).

- Get students to participate in active research or development projects around the world by using the Internet. For example, Mike Sublett at Illinois State University is asking his senior geography students to relay their experience of using the Internet to the editors of the *Journal of Geography in Higher Education*.

Tutors may also derive significant stimulation from using ICT:

- Enlivening stale lectures (see Contribution 1).

- Enriching mechanical practical exercises (see Contributions 1, 7 and 8).

- Introducing personal research into teaching (see Contribution 9).

Contribution 4:   Encouraging students to contribute in seminars

Context

Many geographers (for example, Unwin, 1984) have commented on the difficulty of getting students to make oral contributions in tutorials and seminars. Some of the difficulty stems from the students themselves (for example, shyness or lack of preparation), but some is the responsibility of the tutor (for example, insufficient briefing, or over-willingness to step into the breach when silence falls). (See also the companion Guide on 'Small-group Teaching in Geography' by Clark & Wareham, 1998.)

Applications

ICT can be used in several ways to energise tutorials and seminars, most of them requiring a complete reworking of the traditional form of these events.

- Bring a GIS into a seminar on a portable PC, and collaborate with the students in exploring some geographical data. Hand over to your students the responsibility for initiating lines of enquiry and suggesting reasons for patterns revealed on screen. Your own role should be to act largely as a software operator. Only intervene to suggest further lines of enquiry or to encourage alternative explanations (see Contribution 14).

- Get a group of students to play a mainstream computer game before a seminar (for example, explore a selected scenario in *SimCity*), and ask them in the seminar to explain how it can contribute to their understanding of the real world. (In the case of *SimCity*, students could be asked to produce a written critique of the software as a 'serious' attempt to model an urban system.)

- Ask a student to use a simulation programme during a tutorial to investigate a particular human or physical system (for example, a population projection using the GeoSim software). During the tutorial, ask the student to guide the tutor through the software to illustrate their understanding of the system (see Contribution 3).

Contribution 5: Helping to increase the amount of 'active' learning

Context

Passive learning is not only associated with the didactic lecture. It can occur whenever students' minds are not engaged with the subject at hand — even when they are interacting with computer software! (It should be noted in passing that 'interactive' software should not be confused with 'active learning'.) Passive learning does not always help students absorb new ideas or information, because there is little requirement for them to test new material either against what they already know, or against some problem for which the material might be relevant. Active learning requires that a student engages with the material being studied, and is (preferably) challenged to apply this to some new circumstances outside the situation in which the material was first encountered.

Applications

Carefully designed computer software can provide many opportunities for students to engage in active learning.

- Use thematic mapping or GIS software to test geographical hypotheses. This could be done in a conventional practical class, or during a lecture. The lecturer poses a question, displays some mapped data, and then invites students to respond.

- Use a simulation program or tutorial software in follow-on activities to test student understanding of ideas introduced in lectures.

- Use quiz or games programs to help improve student acquisition and retention of geographical information. There is a large literature, particularly in reading studies, on improving student recall of factual information. Quiz programs have been used in the USA to address the national concern with students' lack of basic geographical knowledge.

- Recast the use of stand-alone, computer-based learning activities in the context of Kolb's experiential learning cycle. (See Mellor's, 1991 example, and the discussion in Section 1 of this Guide.)

It should not be forgotten that interaction amongst students, as well as between students and staff, can also be very powerful means of inducing active learning.

Contribution 6:   Increasing the realism of practical work

Context

Practical work is often based on small-scale datasets, partly because this makes them easier to handle by hand, and partly because textbooks do not have room to include extensive tables of data. One of the more important benefits of computer technology is that it enables students to handle large datasets almost as easily as they formerly analysed small datasets. (This is one dimension of the 'emancipatory' role of computers in education.) Large datasets are now readily available from a number of sources: census and other socio-economic data from the ESRC Data Archive at the University of Essex [EDA]; geographical data from MIDAS, at the University of Manchester [MIDAS] and EDINA at Edinburgh University [EDINA]; and diverse data from the Internet. This makes it increasingly difficult to ignore the benefits of using such data in geography courses.

A significant advantage of gaining access to complete enumerations is that students may be able to forgo the need to learn about inferential statistics. Although these methods are essential when dealing with sample data (indicating how to relate sample results to a population), they frequently switch off non-numerate students who are introduced to them in compulsory techniques courses. However, it should not be forgotten that some supposed enumerations suffer from residual bias (as in the case of road traffic accident data), while the use of even enumerated data does not necessarily eradicate the problems caused by modifiable aerial units and the ecological fallacy. (See Contributions 3, 6 and 9.)

Applications

- Use remotely-sensed data to analyse land-use patterns (see Bishop *et al.*, 1995 for several other case studies).

- Use census data to explore a variety of topics in human geography, from poverty and social segregation to urban decentralization and unemployment.

- Use DEMs in physical geography studies of slopes or landscape impacts of new developments. Bruce Gittings at the University of Edinburgh maintains a valuable catalogue of DEM-related data on the Internet [EDEM].

- Visit Web sites that collate diverse information on a particular subject. Hazards are particularly well covered; see, for example, 'The Electronic Volcano' site [EV].

Contribution 7: Connecting students to live information on the world

Context

Practical work in geography traditionally relies on the use of 'packaged' datasets (for example, population censuses) that rapidly become out of date. The availability of environmental monitoring equipment, including data loggers, weather stations and satellite receivers, has radically transformed the currency of data that can be used in routine teaching and learning. In addition, many agencies post regularly-updated environmental and social information on the Internet which can be downloaded free of charge for use in practical work. Data from such sources can often be available for study within moments of their acquisition. Moreover, data are often re-sampled on a regular basis, ranging from a minute (for example, the Houston traffic monitor [HTM], to less than half an hour (for example, METEOSAT images), to several days (for example, LANDSAT and SPOT satellite data).

The primary educational advantages of using current data are that they can:

- improve the up-to-datedness of otherwise 'historical' courses;

- highlight the dynamic nature of many environmental and social phenomena;

- enable the study of environmental phenomena through time;

- illustrate recent events with a geographical dimension (for example, environmental disasters) which might otherwise take months or years to appear in print.

The motivational benefits of using current data should also not be overlooked (see Contribution 3).

Applications

- Use local weather station data in meteorology courses (for example, Perry, 1998; [WW]; [EPAMD]).

- Use images downloaded from satellites for use in practical work to support a wide range of environmental courses (for example, [UNMI]; [DSRS]; [NCDC]).

- Use 'live' Web environmental data, such as urban traffic (as in the Seattle traffic conditions map [STM]), atmospheric pollution (as in the Cambridge air quality monitor [CAQM]), or weather (as in the Macquarie University automatic weather station [MUAWS]).

- Download current information on recent environmental or social events. For example, volcanic eruptions [VE]; earthquakes (e.g. the USGS earthquake site [USE] or the Kobe home page [KE]); social/environmental disasters (e.g. famine and war reports in CNN archives); or areas experiencing environmental pressures (e.g. the Virtual Galapagos Web site [VG]). Phil Gravestock at Cheltenham & Gloucester College of Higher Education visits Web sites for the Montserrat volcano on a daily basis (e.g. [MV]), and prints out relevant material which is displayed on a departmental noticeboard. (John Hunt in the same department used the same approach during the 1997 Kyoto Conference, accessing current information from the New Scientist Web site.) Acting as an information intermediary in this way means that students can have up-to-date information without having to log onto the Net themselves.

- View live pictures of sites around the world to illustrate an area-studies course. Examples of popular locations are the viewer-controlled video camera located on Mount Fuji [MFVC], and the video cameras on the active volcanoes of Mount Etna, Vulcano and Stromboli [MEVC]. An up-to-date list of Web-cam sites around the world is available at [WCS].

Contribution 8:   Getting across difficult concepts and ideas

Context

Many 'technical' topics in geography (such as sampling, or statistics), are often viewed as 'difficult' by students, particularly by those who have a relatively poor level of numeracy. There are also several theoretical and conceptual ideas (such as various aspects of the methodology and philosophy of geography) which many students find difficult. There are many ways in which the use of appropriate software can help students to cope with such techniques and ideas: working through a tutorial program, running a simulation, or undertaking a role-playing game. Sometimes, such activities are best undertaken in preparation for an introductory lecture (perhaps for motivational reasons, as in Contribution 3), sometimes they are best used as reinforcement after other formal classes, and sometimes they are best undertaken instead of conventional teaching. Guidance from tutors will usually be necessary for individual students, or groups of students, to decide which educational strategy is best (see Contribution 3).

Applications

- Use a tutorial program to introduce statistical concepts. General tutorial software includes 'Statistics for the Terrified', and the STEPS Project software, which uses rainfall data to illustrate regression analysis. More specific to geography is the *GeographyCal* 'Social Survey Design' unit, which introduces the concept of hypothesis-testing by means of a sampling exercise.

- Use simulation software to demystify numerical formulae. For example, Mike Tullett at the University of Ulster wrote his own programs to introduce various meteorological concepts (see Case Study 3, Section 1.5, for details). Some of the *GeographyCal* programs could also be used in this way (for example, the

'Weather and Air Quality' unit could be used to introduce meteorological concepts such as lapse rates). Finally, general-purpose spreadsheet software can be used to build sophisticated environmental models (cf. Hardisty *et al.*, 1993).

- Use an environmental simulation program (for example, the SMOKE program embedded in the 'Weather and Air Quality' *GeographyCal* unit) to illustrate the randomness inherent in the behaviour of many environmental systems.

- Use a world modelling program (for example, LIMITS or IFS) to illustrate the counter-intuitive behaviour of complex social and environmental systems.

Contribution 9: Linking research to teaching

Context

For some tutors, personal research interests have always been used to inform teaching. In the 1970s, for example, geographers funded by the NDPCAL project to increase the use of IT in teaching adopted the 'trickle down' approach, whereby computer data and software developed in higher education was reworked for use in teaching geography in schools. Nowadays, computers are widely used for geographical research, with the result that research software and data can readily be adopted for use in teaching the subject at first and second degree level, and particularly in final-year option courses. Some geography departments (for example, at Southampton University) base much of their undergraduate curriculum on research undertaken by tutors. (See Contributions 3 and 8.)

Applications

- Adapt simulation models developed for geographical research. Useful role models are provided for physical geographers by Kirkby *et al.* (1987), and Hardisty *et al.* (1993).

- Use census data to explore urban social patterns. Having access to the same data as professional researchers makes it possible for (perhaps advanced level) students to redo some of the analyses that underpin research findings. A recent example is the analysis performed by Peach (1996) on 1991 population census data in attempting to decide whether British cities have ghettoes.

- Use the wealth of climatological data available on the Internet for student practical work (Perry, 1998).

See also Contribution 21, which discusses the potential of using the expertise of researchers located at universities elsewhere in the world.

Contribution 10: Speeding up the learning process

Context

We have already seen (for example, in Contribution 8) how computers make it possible for students to analyse large datasets, and thus engage in the exploration of real geographical problems. However, delays frequently bedevil student analysis, and can undermine their enthusiasm for investigative work (see Contribution 3). There are several dimensions to this problem:

- In learning contexts where students gather their own data (for example, on field courses or during dissertation preparation), there is often a time lag between data acquisition and analysis. This can mean that there is a critical time gap between a student coming up with an hypothesis and the opportunity for testing it. Some published studies in interactive computing suggest that there should be a maximum delay of about 3 seconds before software responds to a user's input. A gap of hours, days or weeks between having an idea and being able to explore it can have a significantly negative impact on the development of the student as an active learner, and maybe a negative attitude towards engaging in research activities.

- In some geography departments, there are simply too few computers available to permit simultaneous access to relevant analytical or tutorial software by all students enrolled on a course, and this leads to inevitable delays in carrying out follow-up work (see also Contribution 19). The Dearing Report (NCIHE, 1997) suggested a solution to this problem in the Student Portable Computer, but few departments have the resources to provide enough machines for all their students.

- Students have to undertake preliminary processing of relevant data in order to convert them into a suitable form for effective analysis. Examples include the pre-processing of raw satellite data to remove data errors (thankfully, this is usually done before data are acquired from most sources), and the entry of questionnaire survey data from paper forms onto spreadsheet, database or statistical software.

There are several contexts in which data exploration is an important element of the student's learning process, and notably fieldcourses and dissertation preparation (see the companion Guide on 'Fieldwork and Dissertations in Geography' by Livingstone *et al.*, 1998). In these areas, the ready availability of computer power at the place and time where primary data are gathered, can contribute significantly to the successful prosecution of an investigation. There are several advantages to be gained:

- time saved in undertaking the research, because students do not have to get back up to speed after an enforced 'lay-off', and reduced necessity for revisiting field locations to gather additional data later deemed essential;

- improved ability to explore data, particularly rapid adjustment of research directions in the field on the basis of feedback from immediate analysis;

- development of a positive attitude towards engaging in research activities.

Applications

ICT can be used in several ways to remove the gap frequently interposed between data gathering and data analysis:

- Take a PC on field courses, entering and analysing data in the evening so that the following day's activities can be planned on the basis of the results (Gardiner & Unwin, 1986).

- If heavy-duty analysis is required, or appropriate software is only available on a central or networked computers at the university, connect a data logger or PC available out in the field to computer resources back on campus over telephone lines, using an acoustic coupler or modem. Transfer data to the campus facilities for analysis, and download the results for subsequent interpretation in the field.

- Take an integrated data logger into the field for live environmental monitoring and analysis. In the US, school children have used hand-held PCs (Apple Newtons) to undertake environmental projects in their own locality.

- Issue students with a hand-held PC for carrying out dissertation-related surveys in the field. For example, this could be used to enter social survey information directly into data processing software. This would not only remove the paper form-filling stage of data gathering, but would also make possible preliminary analyses that could be used to guide subsequent data gathering.

- Pre-process geographical data, or acquire pre-processed versions of data (for example, from appropriate suppliers, or on the Internet) so that students can be up and running immediately with their own analyses.

- Re-organise computer-based practical work on a small-group basis, rather than individually, so that students can begin their analytical work more rapidly even if there are a limited number of PCs available.

Contribution 11:   Handling large(r) numbers of students

Context

The recent introduction of a mass higher education system in the UK has led to some educators advocating the use of the mass lecture as the only realistic — and efficient — solution to the resulting problem of engorged enrolments. As we have seen in Contribution 1, ICT can help out when using traditional teaching methods such as lectures, by introducing an element of variety into the teaching process. More radically, tutorial-style software can be used instead of traditional didactic methods to provide basic instruction and testing on introductory topics. However, it must be recognised that this approach requires that sufficient numbers of PCs be available, or that student access to limited numbers of computers be carefully scheduled during the course of the academic year.

Applications

- Introduce first-year students to the essential ideas and skills of geography using appropriate tutorial software. This approach is described in Case Study 1 (Section 1.3). Several of the TLTP *GeographyCal* modules (for example, T20 on 'Map Design') were specifically designed to introduce students to the essentials of technical topics in large networked PC labs.

- If there are too many students in the department to permit regular face-to-face access to tutors, then it may be more effective to use the college network (and/or the Internet) to enable students to communicate with their course tutors on-line (see Contribution 12).

- Set up a departmental 'help line' on the college network, perhaps based on the college email system, so that students can find out essential information from, or report personal problems to, a course secretary or academic 'duty tutor' (see Contribution 20).

- Permit students to work at home and/or hall of residence using dial-up connections (see also Contribution 20), so that pressure is taken off scarce campus resources.

Contribution 12: Improving student access to tutors

Context

One of the most significant changes in British higher education during the early 1990s has been the rapid increase in student numbers enrolled in many geography departments. Although several new problems have emerged in the train of this development (see Contribution 11), several existing problems have got worse. One of the most significant of these is the frequent inability of students to discuss study problems with their tutors as and when they arise. To the absence of tutors who are engaged in active research is now added the difficulty of very large numbers of students attempting to see their tutors at potentially the same time, in order to discuss lecture ideas, tutorial assignments, fieldwork follow-up or dissertation preparation.

Perhaps the most valuable assistance that ICT can offer in this situation is through improved communications, using internal networks and/or the global Internet.

Applications

- Use email for student-staff communication. At the University of Leicester all first-year students in the core practical geography class send an email message to a selected member of staff as their first network familiarisation task. (For further details of using an intranet to encourage student communication, see Case Study 6 by Christine Steeples *et al.* at the University of Lancaster, Section 3.3.3, and Case Study 7 by John Stainfield at the University of Plymouth, Section 3.3.3.) At the Open University, where students live and study off campus, most contact is still by telephone. However, an increasing number of students are now using email, and steps are afoot to transfer more study activities to the Internet.

- Encourage students to use the Internet to communicate with experts elsewhere in the world who will answer questions about matters related to their own area of expertise, or who may be willing to help students who are looking for research project ideas. Examples are provided in Contribution 21.

- Encourage students to submit comments while a module or course is under way, thereby providing valuable formative feedback (see also Contribution 25). In the commercial world of games software, many companies (for example, ID software, the company which produces the well-known DOOM and QUAKE games), make a point of answering all email messages sent in by players, thereby acquiring valuable feedback from expert users of their products.

Contribution 13:   Enabling collaborative learning

Context

Interaction between students on campus-based courses has traditionally been restricted to face-to-face, group activities such as tutorials, seminars and practicals, while off-campus collaborative work is usually confined to group fieldwork exercises. (By contrast, most dissertation work is undertaken privately by individual students, as are the projects carried out during work experience periods and industrial placements.) Given that there is mounting evidence (for example, from employers' surveys) of the importance of team-work in the job market, further opportunities to develop collaborative working in geography courses ought to be encouraged.

Applications

ICT can help to develop group-work capabilities. In particular, there has been a rapid increase in 'network smart' computer software in recent years, including Electronic Mail, the Web, 'groupware', audio- and video-conferencing, electronic meetings and media rooms. These facilities can be used in several ways to enable students to undertake joint work, whether in pairs or in larger groups. Here are two styles of interaction:

- *Competitive interaction*. Perhaps the best commercial examples of the competitive use of networks are the DOOM and QUAKE 'death-match' style games, which can be played by up to 16 players on a local network or across the Internet. One tutor on a Computer Information Systems course at the US Military Academy has adapted the DOOM II game to familiarise students with their tutors and with basic computer knowledge [DOOM]. Although the game currently involves individual students pitting themselves against the computer, it would not be difficult to adapt it further to enable students to compete with one another across a network.

- *Collaborative interaction*. A number of commercial programs (such as Lotus Notes) provide 'groupware' facilities for people working in larger organisations. Some of these permit off-site workers to connect to the organisation, whether they are travelling sales staff or home-workers. One educational application might be an electronically mediated version of the student writing groups described by Hay & Delaney (1994). A more ambitious goal would be the development of what Kouzes *et al.* (1996) call 'collaboratories': dispersed scientific investigations on the Internet. Christine Steeples *et al.* (in Case Study 6, Section 3.3.3) describe the use of networking facilities to develop a 'Learning Academy' in which students undertake a variety of collaborative learning activities.

There are a number of educational innovations waiting to happen that make innovative use of networking facilities. Why not be the first to try them out in geography? Here are three suggestions to get you thinking:

- *Student collaboration on investigative projects*

  One of the major educational opportunities opened up by the Internet is the potential for students collecting and analysing local environmental or social data and making them available as part of a nationally or globally coordinated survey. An example of this approach is provided by the Global Rivers Environmental Education Network (GREEN), which is currently active in 135 countries [GREEN]. This non-profit organisation sells low-cost water monitoring kit to schools, and provides space on its network server to which students can post the data collected in their own river catchments. Other examples include the Global Learning and Observations to Benefit the Environment (GLOBE) program, which coordinates the collection and reporting of local environmental data by thousands of schools in over 50 countries [GLOBE], the 'one sky, many voices' project, which involves students in inquiry-based programmes in science [OSMV], and the National Geographic Society, which is using the Web to carry out its global survey 2000 [NGSS]. On a more local scale, the MetNet Europe project has recently been set up to enable schools to gather and analyse weather data across Europe [MNE], while in Australia schoolchildren are le to collaborate in weather investigations through Project Atmosphere Australia Online [PAAO].

- *Departmental provision of surrogate field trips*

  Fieldcourses are under severe pressure because of constraints on departmental budgets, and there are a number of initiatives under way to harness ICT to reduce unnecessary expenditure, or to improve educational returns from this form of study. One way of using the Internet would be for one geography department to run its field course on behalf of other geography departments, relaying information from the field to other participants through the Web. There should be considerable opportunity for such tie-ups, because many departments use the same field study locations, at home and overseas, and many departments study common issues in the field.

  The only kit required would be a portable PC, a digital camera and/or video-recorder, facilities for editing Web pages, and a means of relaying the pages back to the Web site at regular intervals. (This last step could be done once a day from a field-course base, or it could be done more regularly from the field if GSM telephone or satellite uplink facilities were available.) Indeed, the information flow could be made two-way, because the stay-at-home students could send requests to the field team to gather additional information, or ask the day's guest speaker specific questions.

  For an example of how this might work, take a look at the following two examples which operate in the schools sector: GlobaLearn [GL], which operates remote expeditions; and the JASON Project [JP], which runs an annual participative expedition, and also has remote cameras in the Yellowstone National Park (USA) and Heimaey (Iceland), whose live pictures are made available via the Internet for use in schools.

- *Computer collaboration on computational projects*

  Recently, computer enthusiasts worldwide have been giving spare time on their PCs to crack encryption codes [DN] or to identify extra-terrestrial radio signals [SETI]. They download a small item of software from the Internet which runs in the background whenever their PC is switched on, looking for the encryption 'key' or a meaningful signal in a block of data also downloaded from the Internet server. Whenever a user logs onto the Internet, the software automatically sends its results back to the coordinating computer and downloads further test segments of data.

Several kinds of analytical or computational projects in geography could benefit from this kind of communal approach. For example, the unused computational power of PCs in your geography department laboratory could be allocated to a heavy-duty simulation being run by geography students in a less well-endowed department elsewhere in the world. Fertile links could be established between First and Third World geography departments based on such collaborative projects.

Contribution 14:   Helping students to communicate more effectively

Context

Students typically make two kinds of presentation during their courses: the written essay, project report and dissertation, and the verbal contribution at tutorials and seminars. There has been relatively little change in either mode of delivery over the decades, and for purposes of assessment, the former predominates, although there has been a trend in recent years for oral presentations to be assessed on some courses. ICT can help to improve both types of presentation, either by enabling students to create 'electronic essays', or by providing on-screen displays to accompany oral presentations.

There are several advantages to be gained from adopting these new possibilities:

- In a TV/CRT-rich culture, it is myopic to expect students only to produce written reports on paper;

- Electronic essays can benefit from the use of extensive cross-references, in the form of embedded 'hot links' (as will be the case with the Web version of this Guide and the accompanying Resources Database!);

- Using a computer during a presentation can provide a valuable psychological 'prop' for nervous or shy students, and can also help to integrate the use of ICT into mainline study and assessment activities.

There are, of course, serious issues and problems to be overcome when adopting this approach, including those of plagiarism and copyright.

Applications

- Ask students to make a live presentation in a seminar or tutorial, basing (some of) their talk on 'live' analysis of data on (say) a portable PC. Useful software would include a spreadsheet for displaying data as graphs, or a mapping package to display data as thematic maps (see Contribution 4).

- Ask students to 're-version' material already available in electronic form (for example, a CD-ROM encyclopaedia) in order to tell a geographical story, or to argue a particular case.

- Ask students to produce an essay for 'publication' on the Web, with a request for constructive criticism from anyone who reads it. There are large numbers of such 'essays' on the Internet, currently mainly from postgraduate students. For geographical examples, see the Web pages maintained by Jeremy Crampton and others at the Department of Geography, George Mason University [GMU]. This kind of presentation can help students in several ways: they learn how to present a case most effectively for an unseen and perhaps eclectic audience; they learn the difficult art of revising draft ideas; and they learn the importance of building up a global network of experts who can comment on their own particular field. For related discussions, see the companion Guide on 'Transferable Skills and the World of Work' (Chalkley & Harwood, 1998).

Contribution 15:   Coping with non-geography students

Context

The adoption of a modular scheme within universities, allied to an increase in the number of mature-age students entering higher education, has contributed to a growing diversity of student experience of geography in our degree courses. Many first-year students now take modules, or even enrol on an entire course in geography, without the kind of extensive school background in the discipline that was formerly expected of single-honours geography undergraduates. Several conventional strategies have been adopted to cope with this 'problem', including running pre-semester remedial courses, organising a 'buddy' system among students, and including additional tutorial support. ICT can also be mustered to help enculturate non-geography students into the discipline, and bring them rapidly up to speed on basic information, concepts and techniques.

Applications

- Use tutorial software to provide remedial instruction in fundamentals of geography. A useful resource is the *GeographyCal* module on 'Map Design', which was partly designed with a remedial role in mind.

- Use a problem focus which encourages students to see the usefulness of geography as a discipline (see Contribution 3). GIS software can often prove to be an ideal tool to use with students from diverse backgrounds.

- Adopt multi-disciplinary case studies, particularly where students are drawn from a variety of disciplines. GIS-based projects are also particularly suited to cross-disciplinary treatment; reference can readily be made to a variety of prior student experiences, ranging from history and archaeology to business and marketing.

Contribution 16: Coping with slow-lane and fast-track learners

Context

The issue of how best to handle mixed-ability students has surfaced on the higher education agenda, in part because of the conversion of the tertiary sector from an elite to a mass education system, and in part because of the increasing numbers of mature-age students entering the system. 'Better' students, and particularly those with considerable background in geography, often wish to move ahead more rapidly than 'weaker' students. Many mature-age students have rather a low self-esteem in terms of their ability to study, and often like to move ahead more cautiously than younger students.

The ability to provide students with individualised instruction is perhaps one of the oldest justifications for using ICT in education — though it has not always delivered what many originally promised. Nevertheless, computers can provide one-to-one study experiences, and are often available at times that tutors are not.

Applications

Many traditionally taught modules, or elements of modules, can be converted into an independent study mode, using the computer to deliver the instruction (for example, with tutorial software), the associated practical exercises (for example, with data analysis software or simulation models), or both. An example of this approach is provided by the Environmental Monitoring Technologies module at Middlesex University (See Shepherd, 1996, and Case Study 5, Section 1.17).

- The use of computer-managed learning (CML) software can help students with varied learning speeds to adopt a 'just-in-time' learning strategy (see Contribution 23).

Contribution 17: Removing the burden of remedial/basic teaching

Context

Non-geography students (see Contribution 15), weaker students and slower learners (see Contribution 16) pose particular problems for tutors who adopt a 'one-speed' learning strategy. The classic solution to this problem is to offer some form of remedial teaching, either in a specially designed course that precedes the regular geography course, or else as additional tutorials that run in parallel with the geography course. However, both solutions tend to require intensive inputs of tutor time. Computer-delivered remedial material can provide a suitable alternative, with the following advantages:

- It requires little tutor involvement after the initial period of software development, testing and implementation;

- It can be accessed at any time, by any student, wherever there is a networked PC available (this assumes that the remedial materials are loaded onto a university or departmental server);

- The same material can be studied over and over by students, perhaps with some (random) variation in the content determined by the software, until they feel they have reached a suitable level of knowledge and expertise.

Applications

- Use tutorial materials on the Web to bring students up to speed on the essentials of such technologies as remote sensing [RSCC], GIS [GISCC] or the Internet and the Web [NS].

- Use interactive courseware on general Study Skills, such as the 'Making Sense of Information' *GeographyCal* unit. (See also Contribution 1, which describes the department-wide adoption of TLTP *GeographyCal* skills units at the College of St Mark and St John in Plymouth.)

Contribution 18:   Helping part-time learners

Context

The long-time distinction between part-time and full-time students is rapidly becoming a thing of the past. A growing number of universities make no distinction between these two groups of learner; CAT (credit accumulation and transfer) schemes encourage part-time accumulation of credits over lengthy periods of study, and at more than one institution; and many universities encourage forms of distance, open, flexible and lifelong learning that render the part-time/full-time distinction almost meaningless. Nevertheless, from the student's point of view, part-time study can still pose particular problems:

- Not all scheduled course modules are available for them to study;

- They typically travel further to campus on their appointed day(s) to use a library, work with specialist facilities, or consult a tutor;

- They often have to hold down a full-time job while undertaking their study (this applies in particular to postgraduate students).

ICT can be used in several ways to mitigate these difficulties.

Applications

- Provide a selection of course modules in the form of independent study materials, partly or wholly delivered by computer (see Contribution 15). The Open University is perhaps the best example of this approach, though only a small proportion of their course materials is currently computer-based.

- Create computer-based materials that can be taken home by students for follow-up work (see also Contribution 20).

- Develop a course that can be studied largely or entirely in distance-learning mode. An example is the UNIGIS Masters course in GIS based at Manchester Metropolitan University, which is largely based on paper materials, but with practical work involving use of campus-based computer facilities.

- Ensure that students can access staff, other students and on-line help facilities while off campus, for example by allowing them to log in to the university computer network remotely. (See also Contribution 12.)

There are, of course, problems to be overcome in adopting these types of support. For example, it may be costly to provide all part-time students with copies of commercial mapping or GIS software, and it cannot be assumed that all students will have access to suitable PCs of their own at home. One solution to the former problem is to base geographical practical work on generic software that most students will have access to. (An example is provided by the use of the Excel spreadsheet for environmental modelling in Hardisty *et al.*, 1993.)

Contribution 19:   Easing the resource access problem

Context

One of the problems that frequently arises when increases in student numbers are not matched by equivalent increases in educational resources is that students find it difficult to gain access to the materials needed to support their studies. Sometimes this manifests itself in a growing number of students chasing the same number of copies of a key text in the library, or by the introduction of a mandatory booking system for PCs or other geography laboratory equipment. Such problems are not, of course, confined to students, nor are they always due to an absolute lack of resources. Having resources in the wrong place at the wrong time can often be a significant cause of the 'resource access' problem. Nevertheless, an ICT solution may be possible to this problem, whatever its root cause.

Applications

- Put a course-related FAQ (list of answers to 'frequently asked questions') on the college computer network that can be searched by geography students, on or off campus. If these are kept up to date, and address issues of the moment, they will reduce the need for students having to chase after scarce printed course booklets or track down a tutor (see also Contributions 11 and 12).

- Establish a pool of borrowable PCs fitted with network cards that can be used by staff or students to access college-wide network resources from any suitable on-campus or off-campus location (see also Contribution 20). This can often help ease the problem caused by the limited number of seats in a bookable PC room.

- Provide electronic versions of single-copy publications which can be consulted by students from any PC connected to the college network. For example, switch from the printed Geo Abstracts to the computer-searchable Geobase on CD-ROM.

- Refer students to electronic journals which can be browsed from any suitable networked PC. An increasing number of paper journals are being made available in parallel electronic versions (for example, *Journal of Geography in Higher Education, Transactions of the Institute of British Geographers, Third World Quarterly, International Journal of Geographical Information Science, Computers and Geosciences*, and *Environmental Science and Technology*), and several new journals are appearing that are being published only, primarily or equally in electronic versions (for example, *Journal of Geographic Information and Decision Analysis* and *Geographical Systems*). You may need to press your library to acquire the necessary subscriptions to the electronic versions. Check a growing list at the JournalsOnline home page [JOL].

- Where the library has a limited range of journals or book titles in some areas of study, introduce students to on-line searching of bibliographic databases such as the Bath Information and Data Services (BIDS) [BIDS], and staff to the Educational Resources Information Center (ERIC) [ERIC]. Most such services provide abstracts on-line, and several can also provide full documents.

- Point students in the direction of electronic libraries on the Web. An excellent starting point is the World Wide Web Consortium's Virtual Library Project [VLP]; details of the geography component can be found at [VLPG], and the environment component is at [VLPE]. Students should also be directed to other 'jump off' points on the Web, such as the geography information gateway on the CTI Web site at Leicester University [CTIG].

- Put course handouts on the college network (see Contribution 22). This is currently being done by a number of geographers (see, for example, Karl Donert's Remote Sensing course notes [KDRS]). A growing number of geography departments are placing general module descriptions on the Web, with more detailed course handouts and assessment details available only to registered students. A useful example of how departments can collaborate in such schemes is provided by the Virtual Geography Department project based at the University of Texas [VGD]. One of the major advantages of transferring all course information and study materials onto the Web is that both on-campus and off-campus students can have access to the same learning materials, with the result that neither group is penalised because of their chosen mode or locus of study.

- Convert course-related video clips to digital form, and permit students to view these at their convenience on a networked PC. This avoids the need for having large numbers of students meeting together at the same time in the same room to watch a video in often unfavourable conditions. If your digital video clips consume too much disk space for your network manager's liking, you could try writing them to a CD-ROM on an institutional CD-writer. Bear in mind, however, that you will have to abide by relevant copyright restrictions if you go down this route.

Many of these strategies imply additional expenditure (for example, subscription to CD-ROM or on-line databases), but some do not (for example, freeware MPEG viewers are available for displaying digital video). But where resource access problems are not primarily due to shortage of funds, but result from pressure points at key times and places, such expenditure should be considered on its merits.

Contribution 20: Learning from hall, home or the workplace

Context

The traditional form of higher education provision is based on the campus model: students study on a specific campus, and have access to almost all relevant learning resources on that campus. This model inevitably suffers from time constraints (most college buildings are open for fixed times during the day, and are usually closed over the weekend or during holidays). Moreover, for students who wish to study from home, or who are based at home (for example, part-time postgraduates), additional spatial constraints also apply.

To these problems should also be added those due to the competition for finite resources on campus — for example, large numbers of students chasing a limited number of PCs to complete project assignments just before a final deadline expires. Relatively low-tech solutions to these problems have been adopted by organisations specialising in distance learning (for example, the Open University), but ICT solutions are also becoming more widely used.

Applications

- Press the central student support services (for example, library and computing) at your institution to establish an on-line 'help desk' service, perhaps based on the college email system, so that students do not have to traipse across to a fixed location on campus only to find that there is no-one currently available, or that the expert they really need to talk to is on duty later in the day.

- Develop the notion of 'local learning communities', mediated by computer networks, rather than relying on physical presence. (Read Melvin Webber's prescient 1963 paper on 'community without propinquity'.) Such communities should include tutors as well as students, and should be enabled by remote access facilities to college networks. A number of UK geography departments (for example, Kingston University) are experimenting with the use of networking facilities to enable students to study off-campus. In the USA, the 'Virtual Geography Department' hosted by the University of Texas Web site, illustrates one way in which resources can be made available to students for remote learning.

- Use ICT facilities to 'time shift' student work away from fixed learning events such as lectures. Conventional video materials have the advantage of being viewable at flexible times, repeatedly, and pored over in detail. Similar benefits can accrue from using digital video clips or self-paced tutorial software, either in bookable ICT rooms or in the student's own room. For example, John Stainfield makes available his Bihari Farmer simulation program on the Plymouth Geography Web site [JSSM] so that students can take it away to study in their own time.

Contribution 21:   Learning from the world community

Context

All geography departments provide a carefully designed menu of course modules, tailored to suit their view of the discipline, the specialisms and interests of teaching staff, and (more recently) their perception of demand in the student marketplace. No single department can hope to cover the entire spectrum of geography with its courses. Nevertheless, modern developments in telecommunications, and especially the Internet, make it possible, perhaps for the first time, for students to tap into any area of geography from the disciplinary confines of their home department.

Applications

- Encourage students to supplement (or even supplant) formal tuition in their own department by using tutorials posted on the Web (see Contribution 17 for examples). There are numerous 'essays' available on the Internet on geographically relevant topics, from volcanoes and air pollution to on-line communities and poverty. For postgraduate students, there is a growing number of Masters theses posted on the Internet, on a wide range of subjects, including many that are relevant to geography.

- Encourage students to learn about places elsewhere in the world by building up a list of links on world regions or localities that can act as illustrations or case studies of ideas introduced in lectures. There is a growing number of Web sites that document countries, cities, regions and localities around the globe. Simply typing a place name into a Web search engine will bring up several resources to choose from. Another starting point is the Yahoo search engine index to world locations [YWL].

- Encourage students, and particularly postgraduates, to broaden their support base by consulting experts on the Internet in their specialist field of study. An environmental example is provided by the Ask a Vulcanologist site on the Internet [AAV], which is based at the University of Hawaii. Other useful examples include: Ask a Geologist [AAG], Ask a Biologist [AAB], Ask an Expert [AAE] and AskERIC [AE]. A valuable list of on-line experts is being compiled by the Electronic Emissary Project based at the University of Texas, Austin [EEP], though this service concentrates mainly on the school sector, and is not geography specific. There is surely an opportunity here for an enterprising geography department or subject organisation... Finally, several research teams in the discipline post unpublished 'work-in-progress' on the Internet, and some are happy to engage in email exchanges with researchers elsewhere (see also Contribution 9).

One of the problems that can follow increased student use of materials on the Internet is that checking suspected cases of plagiarism in assessed coursework becomes potentially more difficult. Perhaps the only solution is for tutors to become as adept at searching the Web, using its increasingly powerful search engines, in order to identify unacknowledged sources.

Contribution 22:   Improving our environmental responsibility

Context

Geographers are often more responsible environmentally in words than in deeds. Two examples spring to mind: the continued use of paper, and the gradual destruction of field environments. The idea of the 'paperless office' has clearly bypassed most geography departments; course offices, noticeboards and classrooms are awash with course-related handouts. For its part, fieldwork can be progressively destructive of much-visited venues (for example, the Lulworths and Arrans of this world), but it can also take its toll in other ways, such as through the removal of rock specimens. Inventive use of ICT can help to improve geography's environmental track record.

Applications

- Put student briefings (for example, for practical classes or field visits) on the local network rather than providing personal printed versions (see Contribution 2).

- Convert printed course readers to digital form, and maybe integrate them with exercises involving the use of computer software.

- Replace rock thin sections taken from primary sources by digital versions viewable as graphical images on computer screens (see Contribution 2).

There are, of course, several significant problems in becoming more environmentally conscious.

- Converting paper materials to digital form can cost time and money, though this is becoming progressively less of a problem due to the widespread use of word-processing and scanning facilities in most departments;

- Obtaining and paying for copyright clearance for materials included in course readers — though this is a problem already faced by those compiling printed readers;

- The preference of many students for reading from paper documents rather than from a computer screen — most standard computer screens are not well-suited to the display of documents originally designed to be printed on paper;

- Doubts about the overall environmental friendliness of computer and communication technologies compared with the printed page.

Contribution 23:   Managing the learning process

Context

An increasingly common feature of many geography departments is the large number of students with contrasting prior experience of geography, different learning speeds and styles (see Contribution 16), following a variety of undergraduate programmes. Consequently, the need to manage individual student learning paths has probably never been greater. To some extent, this management problem is one that must be shouldered by the students themselves, but departments and courses also need to consider how best to provide a flexible learning experience for large numbers of diverse students. Some kind of formal curriculum management system may therefore be required. The advantages of using a software-based solution are considerable:

- Because most aspects of course management can be automated, little additional administrative support is required;

- The system can be made accessible to both students and staff on a departmental or university network;

- The system can be applied across courses, departments and campuses, a boon where students compile courses on a 'pick and mix' basis from a number of departments.

Applications

For some heads of geography departments, the educational management roles of ICT may well become at least as important as the more obvious roles in teaching and learning. Potential applications include:

- Registering students for course modules;

- Monitoring student use of study materials (for example, print resources checked out of an resource-based learning study centre, or number of accesses made to units in a Computer-based tutorial);

- Administering the collection and reporting of module evaluation feedback from students;

- Undertaking routine assessment, especially by means of objective tests (see Contribution 24);

- Providing booking facilities for student access to rationed resources (for example, laboratory equipment, computers for independent practical work, tutorials, clinics or troubleshooting sessions). This can be done by making available a single, stand-alone PC at a key point in a department, dedicated to taking specific laboratory bookings, or by providing students with access to relevant booking software on a network. The latter is more flexible, particularly where students take modules in several departments, buildings, and/or campuses. For students studying off-campus, Web-based booking facilities would be ideal.

- Managing student selection and study of instructional units. One approach is to use software which monitors students' progress in a particular learning unit and then uses this information to decide their next learning activity. This decision may be based on the time taken to undertake a previous learning unit, the assessed mark they were awarded, their perceived need for remedial exercises, or some other criteria. None, some or all of the actual learning activities may be delivered on the computer. For an example, see Robinson (1993).

Contribution 24: Lightening the assessment load

Context

Several problems beset assessment in geography:

- It can be onerous for both staff and students;

- Students are subjected largely to summative assessment, while receiving little benefit from under-used methods of formative assessment.

Several solutions to these problems are explored in the companion Guide on Assessment in Geography (Bradford & O'Connell, 1998). Technological solutions to both these problems are considered below.

Applications

- Insist that students word process all assessed work after an initial easing-in period in the first year. This has the dual benefit of improving students' general skills and making it considerably easier for tutors to mark assignments.

- Where appropriate, reduce the amount of essay-type assessment, and replace this with carefully designed, computer-marked objective tests (Stainfield, 1994). Such tests can be compiled and administered locally, by purchasing PC software (such as QuestionMark). Alternatively, you can use freely available tools to devise your own Web-based assessments (for example, the CASTLE project [CP]), or you can use an on-line test service (for example, the DIADS/TRIADS system under development with TLTP funds at the University of Derby). Don Mackenzie (D.Mackenzie@derby.ac.uk) at the University of Derby can be contacted for details about the DIADS system, and a geographical application is reported by Pearson *et al.* (1997). The TRIADS system [TRIADS] is a collaborative project between the University of Liverpool, University of Derby and the Open University. Alan Boyle at the University of Liverpool can be contacted for details about the TRIADS project (apboyle@liv.ac.uk). See also the set of four papers on optical mark reader (OMR) technology in the *Journal of Geography in Higher Education* (1997), Vol. 21(1), and Weaver (1997). A software developer's rationale for using computer-based tests can be found in [CBT], and guidance on setting effective objective tests is provided in Heard *et al.* (1997).

- Adopt computer-based facilities for collating, analysing and viewing marks and grades. Facilities for collating student marks and calculating grades are now used in most institutions of higher education, some based on departmental spreadsheets, others part of a university-wide student information system. However, there is still a need in many institutions for on-line access to marks to be made available to tutors at departmental level, and for facilities to enable retrospective searches and longitudinal analyses to be undertaken on assessment data.

- Link computer-based assessment to computer-delivered courseware, so that remedial feedback is provided throughout the learning process. Summative assessment may then be based on similar tests to the formative assessment, or else might adopt conventional (for example, essay-type) methods.

Such innovations have resource implications, including:

- Provision of remedial or introductory training in word processing for those that require it (for example, through college computer services);

- Investment in OMR equipment and associated software;

- Training tutors in the skills of composing objective tests.

Contribution 25:   Streamlining course evaluation

Context

Many geography departments require formal student feedback on course modules, usually through a standard course evaluation questionnaire. These can, however, be time-consuming for students (in the time taken to complete them — a chore that is not universally welcomed), and for tutors (in the time taken to analyse and report the results). Some form of savings are possible by careful use of ICT, though perhaps these are biased more towards tutors than their students.

Applications

- Require students to complete evaluation questionnaires using computer software, which then automatically produces a summary of results for departmental use. (See Drummond, 1997.)

- Develop a feedback response facility that can be called up by students using a 'hot-key' combination when they are using any course-related computer software. This would not only enable feedback to be received by course tutors that can be used to improve computer-delivered forms of study, but it could also be broadened to include comments on the module on which particular courseware was being used as a learning resource.

Focus on technology

"The computer revolution of earlier decades has now turned into a communications revolution."

(Starr, 1996, p.50)

This section of the Guide takes a closer look at three technologies that have emerged recently, and which are claimed to provide significant benefits for teaching and learning across the curriculum. The first (multimedia) represents the IT element of this Guide; the second (the Internet) represents the CT element of this Guide; and the third (the World Wide Web) represents a marriage of the two. We will attempt to lay bear the essential characteristics of these technologies, sparing no blushes in the process, and we will identify some positive applications for teaching geography.

In the near future, additional technologies will inevitably appear, and others that are currently in their infancy will mature. Some existing technologies (for example, desktop video-conferencing over the Internet, and distributed virtual environments) may provide support for teaching and learning geography. (For a valuable preview of the educational potential of video-conferencing, see Coventry, 1997.) As for the unknown future technologies, we would encourage you to monitor the technological landscape on a regular basis, perhaps with colleagues and subject-based groups, or perhaps by attending conferences and exhibitions (the first Virtual Reality in Education and Training conference was held in mid-1997), and evaluate new resources for their educational potential. We hope that the approach taken in this section provides some guidance on how to do this.

This section has been written primarily for those are relatively new to ICT. Nevertheless, some of it may be of interest to those who are already well versed in ICT, but who might benefit from a critical perspective on recent technological developments.

3.1 Multimedia

"The addition of high-quality graphics, audio, and video to text, and more powerful editing and authoring software, provide a major enhancement of computer-based learning... Those countries that harness the power of multi-media communications for education and training purposes will be the economic powerhouses of the 21st century."

(Bates, 1994, pp.3-8)

"Multimedia... invites a fascination with technical virtuosity and surface effects that can become a distraction from learning."

(Starr, 1996, p.53)

3.1.1   Multimedia seems to be everywhere! — Exactly what is it?

Multimedia has had an upbeat press in the past few years. As a popular buzz word, it makes a regular appearance in educational circles, where the following products are being developed and used:

- multimedia atlases

- multimedia encyclopaedias

- multimedia tutorials

- multimedia games

- multimedia simulations

Multimedia refers to computer materials that include various types of information, including text, numbers, images, video and sound (Ambron & Hooper, 1988; Nielsen, 1990b). Particularly with the widespread use of CD-ROMs to disseminate digital information, and with information on the World Wide Web moving rapidly from plain text to multimedia documents, multimedia is a technology whose time in education finally appears to have come (Beattie *et al.*, 1994). In recent years, there has been a flood of so-called educational products claiming to enhance student learning because they incorporate 'multimedia', 'interactive multimedia' or 'hypermedia'. Many commercial organisations with expertise in producing multimedia products for entertainment are emphasising the 'learning' benefits of their offerings, particularly in the domestic marketplace. Here, for example, is Microsoft's sales pitch for its multimedia encyclopaedia, Encarta:

> *'It brings learning to life with words, images, animations, and sounds that work together to create a fascinating universe of knowledge... sparks curiosity, opens the door to wonder, and starts a learning experience that never ends... helps you find the joy of learning.'*

In order to appreciate what multimedia has to offer educationally, it is essential that we understand exactly what it means. Unfortunately, the term 'multimedia' is perhaps the worst (or best!) example of a misnomer ever invented, and this goes way beyond its tautological combination of 'multi' and 'media'.

- Multimedia has been around for a long time, well before the advent of computers. All videos, films, books, magazines and other publications which mix text and pictures are multimedia products. Even the Open University is recognised as being "a multimedia institution since its inception" (Greenberg, 1994, p.21), despite the fact that its study materials still consist mainly of non-integrated print, video and audio.

- The terms 'medium' and 'media' are conventionally used to refer to 'channels' or 'instruments' of mediated communication. Thus, for example, we have such media as the publishing industry, the press, radio and TV, and we also have the media used by artists: paint, ceramics, film, and so on. However, the word 'media' in multimedia does not refer to channels at all — it is not, for example, meant to distinguish between alternative means of information delivery, such as floppy disks,

CD-ROMs, local area networks or the Internet. Rather, it is used to describe the different kinds of information that are bound together in a multimedia product. As it happens, the key feature of modern multimedia is that it comes along a single, digital channel, thereby avoiding all the problems faced by information users in the past who have had to 'cut and paste' information from a variety of sources and forms. Although 'digital multi-information' may be a more precise term for such materials, it has far less of a ring to it than 'multimedia'.

- In much of the educational literature, it is implicitly assumed that multimedia is a cut above other forms of information, and that it almost guarantees effective learning when used by students. This conflation of the technology with supposed educational effectiveness is apparent in Wilson & Tally's (1991) description of multimedia as a "set of teaching and learning tools", and in Rieber's (1994) reference to multimedia as "integrated instructional systems". Indeed, much of the educational technology literature now refers to 'multimedia' rather than to 'computer assisted learning', and computer-based learning resources are being increasingly issued as 'multimedia software' rather than as simulation programs, tutorial software, databases and the like.

Multimedia has generated excitement because of the way it has added value to previous information technology, not necessarily because it is a major advance in its own right. In some ways, the fuss currently being made over multimedia echoes the excitement caused when sound was first added to movies, and when colour was first added to television. Today, the hype is focused on the combination of information in a single digital medium; a major advance, certainly, but not perhaps worthy of a confusing neologism. The 'M' word will probably have a relatively short life, receding from public view as multiple types of information becomes the norm in all digital products, educational as well as secular.

In order to reap the most from multimedia, it is important to see it for what it is: the integration of diverse forms of information in a digital form that increasingly pervades all educational computer materials, but which is not in itself a type of educational product. In order to arrive at an informed decision as to the efficacy of geographical materials involving multimedia, it is important that you ask yourself some level-headed questions before getting drawn into the technology (Olson, 1997). We will return to these issues when we discuss the Internet, which is fast becoming the best-known delivery medium for multimedia.

3.1.2 What are the main educational uses of multimedia?

Multimedia finds its way into two kinds of educational resource: tutorial-style software, and databases.

- *Tutorial software* is designed primarily to teach students particular subject matter. In geography, a growing number of courseware units are being designed, and most now incorporate varied information. The better examples also provide a far more varied learning environment than the crude programmed learning software of a generation ago. An early example of effective use of multimedia is the GIST system (Raper & Green, 1989), which was designed as an interactive demonstrator for basic GIS concepts, and includes several animated explanations of GIS techniques and concepts. Several of the recently completed TLTP geography modules incorporate data analysis, mapping, simulation and other interactive exercises to encourage active student participation in the learning process. It could be argued that in such software, it is the rich set of opportunities for student interaction, rather than the multimedia nature of the information they contain, that is of greatest educational importance. Such interaction includes: intervening at any point in a dialogue; changing a setting in a data mapping exercise; running an animation with varying starting values; suppressing selected features in the information being displayed; query the result provided by the software; replying to self-assessed tests; and so on.

Interaction permits a student to experiment with a given demonstration, stopping it, restarting it, changing parameters, and generally converting the experience into a 'what if?' experiment. Here, we find the benefits of the conventional simulation program (which has been around in geography for at least a quarter of a century) embedded in, and enriched by, a hypermedia environment.

Self-assessed questions and other forms of mini-exercise enable students to test their understanding of a hypermedia learning unit. The best such tests not only provide additional feedback if the student runs into difficulties, but also provide a summative report on their progress. A great deal of thought goes into the interaction with the student, so that the feedback is at least as beneficial as the tuition provided by the main learning sequences.

Nevertheless, it is important to recognise that student-computer interactions in most current multimedia software are ultimately relatively limited, and fall considerably short of the free-ranging discourse that occurs between students and their peers and between students and tutors. Indeed, some recent studies suggest that the styles of interaction currently adopted by educational multimedia actually encourages a reactive style of response by students, who tend to limit their curiosity because of an inability to ask questions of concern to them.

- *Resource databases* are designed for consultation by students when engaged in various learning activities. A growing number of multimedia CD-ROMs have appeared, increasingly from commercial sources (notably CD 'atlases' and 'encyclopedias'), that provide a bank of text, factual information in tabular form, photos, moving images and sounds which can be searched, selected and printed out by students. Information-rich organisations are beginning to capitalise on their resources by releasing educationally-targeted CD-ROMs, in the earth sciences (for example, the United States Geological Survey–USGS), as well as geography (for example, National Geographic and Encyclopaedia Britannica). An increasing number of academics are now collaborating with non-academic organisations (for example, Ferrigno & Wiltshire, 1994; Messing & McLachlan, 1994) or else transcribing archival material (for example, O'Day, 1997) in order to develop multimedia resources for educational use. With most of these resources, the style of educational use (teaching, browsing, asking questions) is left largely to the teacher, although most build in certain student-friendly aids, such as overviews and hyperlinks.

> This kind of multimedia resource is extremely valuable for introducing students to a variety of perspectives on world features and events, and provides materials for use in essays and project reports. However, there is a danger that, without careful tutor guidance, such materials may become simply an 'easy option' for weaker students who see in them a short-cut to filling space in their coursework assignments. There are a growing number of 'armchair traveller' type of CD-ROMs, often produced for the home market, which provide relatively superficial descriptions of localities and events in various parts of the world. If students are to be able to discriminate between useful and worthless information, and derive learning benefits from this kind of material, then tutors will need to provide students with careful guidance in their use.

It is probably futile trying to decide which is the 'better' of these two uses of multimedia. Indeed, opinion is divided between those who see the strength of multimedia as being an educationally neutral form of resource, in which multimedia information is provided in the form of a database, encyclopaedia or dictionary, with the tutor providing the educational context, and those who see the benefits of multimedia being realised by embedding it in interactive, tutorial courseware, in which students experience a multi-sensory engagement with varied information through carefully designed instructional software. The most effective approach probably lies in marrying both uses of multimedia, as demonstrated by The Geographer's Craft project at the University of Texas (Foote, 1994).

3.1.3   Why is multimedia supposed to be so educationally effective?

The benefits of multimedia as an educational tool in geography are assumed to derive from its two main characteristics:

- It includes diverse types of information (text, numbers, pictures, video, sound).

- All information is in a common digital form.

Let us take a closer look at the claims made for the educational benefits of these attributes.

Graphical information

Geography, so the argument goes, is an inherently visual subject — witness the wealth of geographical images in the form of maps, photos, film, etc. Hence the importance, for example, of graphicacy as one of the discipline's key skills. Consequently, any medium that delivers images as well as text must be an advantage to students learning the subject. The power of computer graphics for teaching elements of geography is widely accepted (Batty, 1985), and a number of recent multimedia products (for example, the GIST 'tutor') blend graphics effectively with text to enhance the learning experience. And yet there is a danger that multimedia will simply lead to the creation and dissemination of too many journalistic travelogues and superficial regional geographies, rather than thought-provoking exercises in which the images are put to better educational effect.

Graphics may be a good thing, but interaction is better. What this means is that we should perhaps see graphics as an outcome of student activity (for example, through the mapping of data) rather than as an input to student learning. In other words, while simple multimedia can induce the same kind of passivity amongst learners as TV and video, well designed interactive multimedia can stimulate active learning.

All-digital information

Digital multimedia scores heavily over conventional audio-visual aids when it is important for students to be able to switch between different sources of information, to compare them, analyse them, cut-and-paste from one form to another, and generally to interact seamlessly with a range of information on a single desktop. For tutors, it is equally valuable to be able to bring together a range of teaching materials in a single digital environment, without having to manually cut and paste them for student use. (Many resource-based study packs even now consist of a motley collection of paper handouts, audio tapes, video cassettes, maps, computer databases, and other sundry items, each stubbornly independent of the other.)

However, there is a third characteristic which is not inherent in multimedia, but which is associated with most current multimedia products:

- The non-sequential arrangement of information.

Hypermedia

In most multimedia products, chunks of information are associated with one another by means of 'links' (or hyperlinks) embedded in the material. These links, which are usually added to the material by the developer, enable users to navigate through the information in a quasi-random way, without being constrained by what someone has called 'the tyranny of sequential information'. This interconnecting of units of information has led to the use of the term 'hypermedia' to refer to multimedia which contains embedded links. A decade ago, when most computer-delivered information was textual, the original term was 'hypertext'.

Where the material contained in a multimedia package is organised in a non-linear sequence, students can take their own paths through the information, building up an individual learning experience that follows from their natural curiosity or a specific problem-solving exercise. This style of material is referred to as 'hypertext' or, where multiple types of information are involved, 'hypermedia' — 'hyper' referring in both cases to the network-like interconnections between basic units of information.

Hypermedia usually goes some way beyond simpler forms of multi-way branching characteristic of early programmed learning materials, where students took decisions at multi-way decision points or selected options from menus. Hypermedia materials generally include 'hot links' between items of related information, an idea that has recently been taken up in the links between Web pages (see Section 3.4). This flexible form of interconnection between multimedia information is supposed to offer major educational advantages. Not only, so it is claimed, can the learner skip elements they are already familiar with, but they are free to pursue their own lines of investigation and learning — in educational jargon, learner control is maximised.

In practice, however, there is a down side to the inter-linking commonly found in hypermedia. Many learners get disoriented or confused by the complexity of the interconnections between the chunks of information they are exploring, and a number of research studies have investigated how and when users get 'lost in hyperspace'. Current thinking is that restrictions be placed on links in educational hypermedia, and that clearer navigational aids and overviews be provided for students (for example, indexes, 'route maps' and quizzes). (See Weyer & Borning, 1985; Hammond, 1989; Hammond & Allinson, 1987; Hammond & Allinson, 1989; Nielsen, 1990a.)

In an earlier section we discussed the problem of constraining students' learning by providing them with inflexible computer facilities. However, we seem to have come full circle, because with hypermedia (which includes the World Wide Web, of which more later) it is possible to undermine student learning by giving them infinitely complex information and study tools. We may therefore have to accept that complete freedom is not necessarily a good thing where digital learning resources are concerned. As an alternative, Elsom-Cook (1988) suggests that students would benefit from using tutoring systems in which they can move back and forth along a scale from totally constrained guidance at one end to total freedom at the other, as the occasion and student requirements dictate.

3.1.4 How does multimedia rate in comparison with conventional teaching materials?

A fundamental question concerning multimedia is whether it takes the student any further than, say, the modern printed textbook. In relation to conventional printed texts, it ought to be made clear at the outset that although there have been numerous studies of the learning effects of combining information of different types in textbooks and other educational materials (for example, Purnell & Solman, 1991), there are no hard and fast learning benefits to be obtained from mixing information on the printed page. So, why should digital multimedia fare any better? Some multimedia is almost identical to printed texts, consisting mainly of text, with a variety of accompanying illustrations — for example, tables, maps, diagrams, photos. To be sure, some multimedia also includes moving images (video clips or animations), and sound (natural sounds, music, recorded speech and commentary). Educationally, however, multimedia enjoys largely the same benefits and suffers most of the disadvantages of conventional textbooks.

In some respects, multimedia is decidedly poorer than its paper counterpart. This is especially the case with screen readability. Not only is the typical computer screen limited in the amount of information it can display, but students find reading large amounts of information from screen less easy than from a book. (This is partly to do with screen resolution, partly to do with the unvarying angle of the screen relative to the viewer, and partly to do with the smaller chunks of information on screen which often necessitates additional forward-and-backward referencing, which is not always easy to accomplish.)

Current technological limitations also mean that video clips, where they are included (and this can be problematic because of copyright restrictions), tend to be only very short extracts, fill less than the full screen surface, and are sometimes extremely jerky. We may, however, be on the verge of a technical revolution in digital video, though the cost of acquiring longer, full-screen extracts will be measurable in multiple megabytes of additional data which many hard-pressed institutional network administrators may not be able to mount for general use.

More generally, it can prove extremely problematic to bring multimedia courseware and databases to a mass student audience, because of the scarcity of space on institutional file servers. (The first-phase TLTP geography modules, for example, require some 80 megabytes of storage space.) You may therefore find that stand-alone CD-ROMs or live access to the Web are preferable to a locally networked multimedia resource in your own institution.

3.1.5   Does multimedia enable experiential learning?

One of the areas in which multimedia is expected to provide a cost-effective role is as a substitute for expensive, experiential learning activities, such as laboratory work and geographical field studies. A growing number of simulation programs are providing cost-effective replacements for laboratory science (for example, the Virtual Telescope, the Virtual Microscope, and the simulated human body). There are also a growing number of multimedia products, including CD-ROM courseware and Web sites, which claim to introduce students to various elements of fieldwork, from the preparatory overview of a field venue to the selection of sites for taking soil samples. At least one well-known optical disc (Ecodisc) has provided a range of experiences for students studying field ecology in a field study centre in Southern England (McCormick & Bratt, 1988; Riddle, 1990).

But how far can the range of learning experiences identified for particular field trips be substituted by available multimedia materials? For example, if one values the multi-sensory exposure of students to a rural or urban environment, then the visuals and sounds of current multimedia courseware will provide a sensorily deprived experience for most of them. There is some research evidence that students learn best when their learning is undertaken in a diverse information environment, and that they can absorb and retain information better if multiple senses are engaged during the act of learning. But the engagement of multiple sensory pathways is not the strongest attribute of current multimedia, which still largely engage the eyes and, to a far lesser extent, the ears.

The evidence for the efficacy of multi-sensory information presentation is far from clear-cut. For example, although some research on bi-sensory information presentation supports the view that information processing would be enhanced by combining the visual and auditory senses, other research suggests that the combination of pictures and words can lead to poorer recall performance. Reviewing the evidence, Mayes (1992, p.15) suggests that "in most cases a single medium will be appropriate", and that the usefulness of multi-modal information presentations may not be inherent in multi-sensory engagement, but will depend on such factors as the characteristics of the user and the nature of the information processing task they are involved in. Two educational lessons can be teased out of this evidence for geography. The first is that 'multimedia' should not be confused with 'multi-sensory'; most current multimedia is largely visual. And second, that a concerted research effort is needed to determine which learning activities benefit most from multi-sensory information and which do not. One of the areas that appears to offer most promise is the multi-sensory representation of geographical information (Shepherd, 1995).

For the foreseeable future, even immersion in rich virtual environments cannot hope to provide anything but a poor imitation of the real world which students can feel, smell and touch. It is perhaps for this reason that developers of so-called 'virtual' field trips and field courses have commendably steered away from attempting to provide an experiential substitute for the real thing. For example, virtual fieldtrips provided on the Web for Malta at Aberystwyth [VFMA] and at Plymouth [VFMP], limit themselves to providing orientation information for students, which can be browsed prior to going on a real field course. For their part, designers of the experimental 'Virtual Field Course' in the UK [VFC] aim to provide data-rich exploratory environments related to real fieldcourse venues, rather than substitutes for the field visits themselves.

Although students can experience a considerable amount of laboratory work in simulated form, they will probably have to wait some time for the multi-sensory technology of distributed virtual environments to mature enough so they can experience simulated field visits that are as rich as the real thing. Until then, there is ample technology to support and enrich fieldwork, without forcing us to engage in the mental gymnastics of deciding whether a fieldcourse is 'real' or 'virtual'.

3.1.6   Is multimedia educationally effective?

It has been suggested in the mainstream educational technology literature that multimedia offers a number of major educational benefits. In particular, multimedia:

- increases the effectiveness of student learning

- increases the efficiency of educational delivery

- increases student motivation

- facilitates active learning

- facilitates experiential learning

- facilitates student-centred learning

Evidence can certainly be found to support most of these contentions, just as there is some evidence for the benefits of other forms of computer-based learning. However, before we get carried away with the idea that multimedia is the best thing to hit education since the invention of printing, it is worth reading the paper by Davies & Crowther (1995) which dismisses each of the claims in this list, regarding them as little more than modern myths. You may want to identify other benefits for this technology when you try it out in your own teaching.

3.2   Communication Technology

"The network is the computer."

(Oracle Corporation)

Everyone involved in the educational process in a geography department can benefit from using internal and external networking facilities. Staff benefit by being able to reach individual students or entire groups of students; students benefit by being able to acquire up-to-date information about their courses, and to get help from tutors and other students when and wherever they need it; and Heads of departments can benefit by being able to market the department and its courses to a wider audience.

The linking of computers into local or wide area networks opens up a number of educational opportunities:

- Collaborative project work by students — for example, carrying out a group project, or writing up a group report. 'Collaborative learning environments' are under active development in a number of disciplines, both in the UK and overseas. In one reported experiment, students around the world participated in a joint problem solving project to explore the problems of water shortage.

- Cooperative or competitive simulations and games. 'Multi-player' is the latest buzzword in mainstream computer gaming. The challenge for geography is to adopt the emerging technology and/or to adapt specific games for educational uses.

Networking of educational resources and activities can be extended to other departments, other institutions and other countries. This can enable a number of activities:

- The international exchange of data and software for teaching purposes.

- Participation in collaborative course design and discussion on problems of course operation. This can be done, for example, through the moderated GeogNet discussion list based at Nene College (Livingstone, 1997) — instructions for joining are given in Section 5.3 of this Guide. In Australia, geography school teachers communicate on issues relating to professional and curriculum development in a similar way [ASCC] and in the US teachers can share educational projects through the Global Schoolnet Foundation [GSNF].

- The searching of 'union' catalogues of educational resources or library collections — for example, the UK's COPAC catalogue, which (at the time of writing) includes bibliographic details of the holdings of two dozen university libraries in the UK [COPAC].

- Provision of on-line lectures or topic summaries.

- Provision of feedback and advice to students on-line, so they do not have to travel to college or fix up appointments with staff in advance.

- Encourage fuller participation by students on contentious or potentially embarrassing topics. Some students find certain geographical subjects identified for seminar discussion difficult to handle, and they fail to participate in face-to-face classes. However, just as some patients will divulge more information to unseen doctors, so it has been found that there is greater student participation in discussions of 'difficult' subjects when email is used (Hall, 1993). In general, network communication can be neutral in terms of race, gender, age, appearance, disability and status in a way that face-to-face communication can never be. As ever, you will have to decide when it is more appropriate to use one or the other.

When broadband (i.e. high capacity) digital networks become available for video-conferencing, this will make possible further educational activities:

- Inviting experts in a particular field to give guest lectures to your students, perhaps on multiple campuses simultaneously, without the inconvenience of physical travel.

- Getting your students to discuss ideas of common interest, or plan collaborative practical work, with student groups in other parts of the world.

- Teachers participating in tele-conferences and distance seminars on aspects of their role as teachers.

The next two sections will explore the specific educational potential of the Internet and the World Wide Web. These are arguably the most significant innovations in communication technology of the past decade and, according to many pundits, threaten to revolutionise the way that higher education is delivered in the future.

3.3   The Internet

"If I were still a teacher I would probably take an axe to any Internet-attached PC that showed up in my classroom."

(Henning, 1997, p.89)

3.3.1   What is the Internet, and how does it differ from our university computer network?

The Internet is a network that connects individual computers and networks of computers across the globe (see Meleis, 1996 for a useful review). Because of its global scope, the Internet can be used by an institution to acquire educational resources produced elsewhere, to deliver course materials to students who wish to study away from campus, and to market its own courses around the world.

However, many of the benefits of the Internet can be, and have been, achieved with internal networks. Laboratory, departmental, campus-wide and university-wide networks have been widely used for some time for a variety of roles in research, administration and teaching. They are used to provide information, to deliver courseware, to facilitate staff and student communication, to organise the acquisition of course feedback, and for various course adminstration and management tasks. Because of the recent popularity of the Internet, a new term ('intranet') is now often used to refer to an organisational network that is used for information sharing exclusively among its members. (To be more precise, intranets usually adopt some form of Web technology, of which more in the next section.)

The Internet hosts a number of distinct services which are of considerable potential value to higher education:

- Email

- Discussion groups or mailing lists

- Newsgroups and bulletin boards

- Text-based chat 'rooms'

- Conferencing (audio or video)

- Access to remote data

- The World Wide Web

Geography researchers, and a growing number of teachers and students, make regular use of email to communicate with collaborators and advisers within and between educational institutions. Groups of staff and students can also share ideas by using newsgroups, bulletin boards, and special-interest discussion lists, usually moderated by a willing enthusiast. There is the additional promise that the Internet will be able to host real-time video-conferencing in the near future.

Although small amounts of information can be exchanged across the Internet using email, special-purpose facilities (using the file transfer protocol or 'ftp') are better suited to accessing files of data at remote locations. The Internet is awash with free data, much of it geographical or environmental, and geography staff are increasingly able to enrich student practical work by downloading such data for analysis and mapping exercises. Bishop *et al.* (1995) provide examples of work undertaken with such data in courses on environmental conservation, remote sensing and geomorphology. Finally, the World Wide Web is a global collection of text and multimedia documents, available on computers connected to the Internet, which are linked to one another to form an infinitely complex information network. In the last few years, the Web has become the fastest-growing and most popular use of the Internet.

3.3.2 Why is the Internet so important educationally?

This question can be answered in two words:

Information   The Internet provides access to a seemingly limitless array of information, stored on the millions of computers connected to the Internet worldwide. It is estimated that there are currently 20 million unique addresses on the Internet, and that the number of documents available is doubling every couple of months.

Communication   The Internet enables students and their tutors to communicate with other students and staff anywhere in the world, as well as with many other kinds of interested party: subject specialists, government and NGO spokespersons, commercial organisations, librarians, interested people in their homes, and others. By connecting people across the globe, the Internet is claimed to provide both staff and students with immediate access to the opinions and views of other learners and experts, thus making collaborative learning a reality.

Because of these characteristics, proponents of the educational value of the Internet suggest that its real significance lies in the way in which it will begin to break down the tyranny of student-teacher and student-campus links, thereby democratising the educational process. Students will no longer be tied to the limited information and staff available at the campus of their choice, but will be free to select whatever information and advice they need from anywhere in the world. Students will no longer feel restrained by the links that persist at traditional universities; the ties between a student and a particular location that have characterised higher education for centuries will be at an end. In sum, it is claimed, education will for the very first time be truly democratised, with students taking their rightful place at the centre of an universal higher education delivery system.

Whether or not we buy into this grand vision, it is clear that several groups of student will benefit from the communication potential of the Internet:

- those without regular face-to-face contact with staff or other students (for example, students enrolled on open or distance-learning courses, or students engaged in work-based learning);

- those with special interests not shared by staff or students at their base institution (this is especially likely with students preparing dissertations on specialist topics);

- those who spend some part of their learning time away from their institutional base (for example, students on a work placement, or studying abroad).

However, two words of caution are necessary here. The first concerns the uneven distribution of PCs and Internet connections throughout the world; the distinction between information 'rich' and information 'poor' students is likely to remain well into the future. Secondly, although the Internet makes global communication and collaboration possible, a significant limit is placed on student and staff participation in real-time discussions across the Internet by their physical location in different time zones. As institutions of higher education reach out to students across the world through Internet-mediated distance learning, the practical difficulty of arranging synchronised communication will inevitably limit the potential benefits to some groups of students. 'Near' and 'far' may be given new meanings in the global educational village, but the friction of time will impose a similar constraint on communication as was previously exerted by the friction of space.

3.3.3 What are the main educational roles of the Internet?

The Internet can help both geography staff and students in a number of ways. For example:

- communicating with other individuals
- discussions with interest groups
- downloading datasets
- acquiring supplementary study materials
- providing course-related information

The first three of these will be discussed in this section; the remaining two will be considered in the next section, because they are increasingly related to the World Wide Web. For complementary perspectives, see Donert (1998) and Livingstone & Shepherd (1997).

Communicating with other individuals

One of the most straightforward educational applications of the Internet involves the use of email to communicate with other people on a one-to-one basis:

- *Student-to-student:* Individual students can keep in touch with each other, especially if they are based on different campuses, or are undertaking group assignments;

- *Student-to-tutor:* Many academics now use email as a major form of communication so it can be a more efficient way of getting responses to straightforward questions or to setting up meetings than trying to catch them in their office or pushing scraps of paper under their door. For distance learners email can be an invaluable link.

- *Students and academics elsewhere:* Because email is so widely used in the academic community, and because answering an email is often easy (technically, anyway), many academics will respond to requests for information by email, whereas they might not go to the trouble of replying to a letter.

- *Tutor-to-tutor:* Organisation of the curriculum is improved if staff consult their email regularly. Changes to teaching arrangements can be rapidly posted, discussed and agreed, and team teaching arrangements rapidly revised. Course tutors can also consult colleagues in other institutions about common teaching problems.

To communicate in these ways requires a personal email 'address' and access to a computer that is connected to the Internet. In most institutions of higher education, staff (and in many cases students) are issued with their own email address by the computer centre, and email facilities can normally be used from any computer in the institution. For those with a computer at home, a dial-up connection to the university network can often be arranged through a telephone (or ISDN) line, and from there onto the Internet, but a modem will be needed to connect the PC to the telephone system. Alternatively, a separate email account can be arranged with a commercial provider.

In some institutions, communication among staff and students is mediated by proprietary software. A popular option at the time of writing is Lotus Notes, which is being used by geographers at Lancaster University (see Vincent *et al.*: 1997, and Case Study 6 in this Section).

Discussions with interest groups

Email also allows the same message to be sent to a number of people simply by typing a list of addresses at the top of the message. This is a very efficient way for course leaders and heads of departments to distribute the same information to groups of colleagues — provided they all read their email regularly. This 'broadcast' approach can also be used by staff to contact students to give them up-to-the-moment information about a particular module, or by a groups of students in order to coordinate group study assignments. (Group distribution lists can easily be set up by individual users with most email software.)

A more formal version of communicating with a group of people is through an email 'discussion list' (also sometimes called a 'mailing list'). Around the world there are thousands of lists which allow individuals to share information on subjects of common interest. The main difference from standard email is that on many lists there is a coordinator (or 'moderator') who receives messages and vets them before sending them out for others on the list to read. There are many discussion lists covering specific areas of research interest to geographers; the premier discussion list in the UK for educational matters relating to geography in higher education is GeogNet, based at Nene College (GeogNet@nene.ac.uk)

— see Section 5.3 of this Guide for joining instructions. A more general one is DeLiberations [DL], which also contains a geography section. If you want to read the information on one of these lists, and maybe contribute to the discussion, it is usually fairly straightforward to join, although exact instructions vary from list to list.

Downloading datasets

Files of information can be downloaded from other computers on the Internet using email, ftp (the File Transfer Protocol) or Web methods (the StatLib system [SL] uses all three!). Some organisations, notably government bodies with a statutory obligation to make information accessible, are providing datasets on the Web or via ftp. The USA is a particularly good source of digital data, because of the right of US citizens to have access to information collected using their tax dollars. Datasets are available particularly for meteorology (including weather satellite images), hydrology and population censuses. For example, the US Department of Agriculture makes available its catchment hydrology data from its Web site [USDA]; the EROS Data Center provides a selection of free satellite images [EDC], and other satellite imagery can be downloaded from the NASA Web site [NASAI]. Such datasets can be used by tutors for practical exercises, and also by students looking for material to be analysed in projects and dissertations.

There is not enough space here to list even a fraction of the possible sources of geographical data on the Internet. (See the next section on the Web for further details.) If you know the name of a data file you are interested in, then you can usually find its location by using the ftp search engine based at the University of Trondheim in Norway [FTPSE]. If your particular file can be found, then this engine will return its complete Internet address. Many Web sites (described in the next section) also include links to datasets that can be acquired through ftp. If you know the full 'address' of a downloadable data file, and it begins with 'ftp:', then email software or a Web browser can usually use the file transfer protocol to log onto the host computer and download the file. Here are some practical tips on downloading data from the Internet:

- Some large Internet files may take a considerable time to download, so it is best to access a remote site at times of the day when they are not busy — night-time at the host site is best;

- It is usually quicker to download large files using the ftp approach than by using the Web's http approach;

- Many data files on the Internet are very large, so make sure you have enough free disk space on your computer before you start downloading them;

- Choose your datasets wisely — many do not have any documentation, and this may make it difficult to use them effectively for teaching purposes;

- Check that the data are not wildly out of date — this could undermine student perception of the Internet as a source of current information.

Case Study 6

Collaborative learning using computer-mediated communications (CMC)

At Lancaster University, a learning technology strategy has been evolved which focuses on the use of ICT to support collaborative learning. The University has been using asynchronous computer-mediated communication to support collaboration in teaching, learning and research since 1988. This approach is based on the belief that much significant learning is essentially communal. Deep-level understanding arises from the conversations, arguments and debates that take place amongst learners, leading to social construction of knowledge and collaborative validation of ideas.

Computer-mediated group processes (using computer-mediated communications (CMC) technologies, or groupware) enable groups of people to explore different perspectives and solve complex problems in flexible ways. This flexibility can be in terms of time and place. As an example, one undergraduate programme will this year replace face-to-face group work with an on-line equivalent, since it was impossible to make the time and place dependent meetings work for all students. They will have a single face-to-face group meeting at the end of the course.

Two recent projects have been led by staff in Information Systems Services and the Centre for Studies in Advanced Learning Technologies (CSALT) in the Department of Educational Research. They have involved teaching staff from the departments of Geography and Economics. The projects have harnessed CMC technology (specifically Lotus Domino via the WWW) specifically for geography-related projects:

The Networking Academy

This involves an electronic debate about the tensions between environmental and developmental concerns in India. For further information see [TNA].

The Internet as a Tool for Collaborative Learning

The main aim of this project was to establish an international student-based learning project around resources found on the Internet. Three volunteer teams of undergraduate geography students at Lancaster University (UK), at University College, Galway (Eire) and at Florida State University, Tallahassee (USA) investigated a common problem by identifying resources available on the Internet and collaborating on their appraisal of these resources. Further details of the project can be found at [ITCC]. For more information about other areas in which this technology is being used at Lancaster, refer to the "Use of Lotus Notes at Lancaster — an overview" in the Paper section of the on-line library at [LNL].

Innovators: Christine Steeples, CSALT, Educational Research; Mark Bryson and Susan Armitage, Learning Technology Support, Information Systems Services, Lancaster University, Lancs. LA1 4YW
Email: c.steeples@lancaster.ac.uk m.bryson@lancaster.ac.uk
s.armitage@lancaster.ac.uk

Case Study 7

Using email to encourage communication among students and staff

At the University of Plymouth, a small-scale experiment in using email and a local Web has recently been completed by John Stainfield in the Department of Geographical Sciences. The idea was to encourage greater communication among staff and students in a third-year module on 'Agriculture and Environment'. Believing that good communication is the key to effective learning, and faced by reduced interaction due (among other things) to the modularisation and semesterisation of courses, John turned to communication technology for a solution.

The main focus of attention on this course was the seminar, which often failed to engage student enthusiasm. In order to improve student participation and commitment, John introduced the use of email facilities as a means of encouraging regular communication between himself and the students, and also as a means of encouraging increased communication amongst students engaged on preparing collaborative coursework. The Web was also used to post course-related materials, for ready access by all students taking the module, and students were also expected to post their individual materials on the email noticeboard so they could be used by other students in the group as part of a group project. In terms of improving staff-to-student communication, the innovation was a considerable success, though it has added a daily half hour of email answering to the tutor's workload. (One payoff is that knocking on the door by students has virtually dried up.) However, as far as inter-student communication is concerned, the benefits have been far less noticeable. Students still have considerable reservations about sharing their work.

This last point is an important one, because it suggests that introducing technology, even for the very best motives, may not always deliver the hoped for — and expected — benefits. Part of the problem in this case has been the baggage of past behaviour that students bring to their final-year studies. But part of the problem is also due to the fact that the communications facilities used by these students were local to the department, and whenever students found material relevant to their project outside the department (for example, in the library), they were loathe to return to the department in order to log onto its network and share the information with others in their group. This problem will hopefully be solved by transferring the email accounts onto the university-wide network.

Innovator: John Stainfield, Department of Geographical Sciences, University of Plymouth.
Email: jstainfield@plymouth.ac.uk

Reference: John Stainfield (1997)

3.4   The World Wide Web

"O what a tangled web we weave,
When first we practise to deceive."

(Sir Walter Scott, Marmion)

3.4.1   What exactly is the World Wide Web?

The World Wide Web, which is also known also as the WWW, W3, or simply the Web, was developed in the late 1980s by European scientists to facilitate research collaboration (see Berners-Lee 1996 for a useful review). The core idea behind the Web is that suitable chunks of information are stored as a standard type of document (adopting the hypertext markup format, or HTML), so that they can be interpreted by any computer connected to the Internet with a suitable 'browser' program. Web documents consist primarily of textual information, with embedded 'tags' indicating how they should be interpreted by the software used to read them. Increasingly, however, Web documents include multimedia content, incorporating images, animated graphics and even sound.

One of the most important features of Web documents is that they can contain the 'addresses' of Web documents stored on other computers linked to the Internet. This means that you can move from one document to another, and from one Web site to another, anywhere in the world, simply by clicking on the relevant address in the current document. It is this network of documents that is referred to as 'the Web'. This should not be confused with the network of computers on which the Web documents are stored, which are linked by the Internet. (The Internet and the Web are both networks: the former is a network of computers and computer networks; the latter is a network of documents.) Navigating through the apparently limitless network of information available on the Internet is often referred to as 'surfing the (Inter)net', though for most people this is more accurately described as surfing the Web.

Many of the benefits of the Web can be obtained without linking to the Internet. For example, many institutions have set up an 'internal' Web, by locating course-related documents on internal computer networks to which only *bona fide* students have access. Such access can either be gained by logging onto the network at a campus PC, or by connecting to the network through a fee-paying arrangement from outside. This has obvious benefits in terms of security and control. (Internal information networks of this kind are sometimes known as 'private Webs'.) In the case of the Clyde Virtual University, which is a service provided for Scottish universities connected to the ClydeNet MAN, lecture course materials are only available to participating institutions [CN]. (See Mulligan, 1997.) The Department of Geography at Leicester University [GDL] keeps most of its detailed course information behind the closed doors of a private Web.

A number of institutions have taken steps to circumvent the delays that increasingly bedevil live use of the Web by students. One of the more effective strategies is to set up a 'cache server', which is an institutional or departmental Web server which stores the Web pages most frequently accessed by students. HEFCE have just embarked on the creation of a second-generation cache server for the higher education community in the UK. These initiatives should help to ensure that students can have rapid access to their Web course materials whatever time they log onto their local computer.

3.4.2   Why is Web information supposed to be so educationally valuable?

There are five characteristics of the information available on the Web that are most frequently cited when justifying its educational use:

Unlimited

We have already discussed the scale of the information available on the Internet; an increasing proportion of this is now being provided as Web documents. It is claimed that students will soon be able to acquire information on any subject from the Web, and that this will solve the problems of their university libraries, some of which are poorly stocked and many of which are severely over-crowded. However, although Web content is growing rapidly, it still represents only a fraction of human knowledge, and little of the paper-based legacy of centuries of publishing has been transferred onto the Web. Unfortunately, many students (and some staff) are treating the Web as their sole port of call when confronted with a study problem.

Free

Unlike textbooks, journals, CD-ROMs and resource-based learning (RBL) packs, most of the information on the Web is freely available. (A good example is The Knowledge Base [KB], which is an online research methods textbook). However, a little discussed question in the higher education community is what will happen when information providers start charging for access to their data. Already, some departments are placing course materials onto password-protected servers, which are only accessible by students who have paid course fees.

Up-to-date

Most printed resources used in education have a relatively long production time, which means that their content is often out of date even before it appears. By contrast, new information can be hosted on the Web as soon as it appears. This advantage lies behind the growing number of electronic journals being published, helps to explain the appeal of on-line 'live' datasets (see Contribution 7), and has stimulated many geography teachers to set up Web sites containing current information to support their courses (for example, Wong, 1998). It also explains the recent appearance of text books accompanied by their own Web sites (for example, Ritter, 1997, [EOWS]).

Multimedia

The Web is rapidly rivalling CD-ROMs as a source of multimedia information. Building on the multimedia capabilities of the Web, many geography lecturers are taking a common approach to the development of educational materials which can be illustrated by a couple of examples. The first is the illustrated slide sequence developed for geoscience students at the University of British Columbia [GUBC]; the other is the Bosnian Virtual Fieldtrip produced for geography students at George Mason University [BVF]. Both provide a series of illustrations accompanied by text, both prompt the student to consider what they see in the accompanying maps and/or photographs, and one of them (the second) poses questions at key points which it expects the student to answer before proceeding.

Interconnected

Information is stored on the Web as documents or 'pages' that are connected by embedded 'links' to other documents or pages. It is claimed that this network (or hypermedia) information structure mimics the way that humans acquire and store information, and thus provides a natural environment in which learners can browse in order to learn. However, as we shall see shortly, there are several problems with the educational assumptions behind this scenario.

3.4.3   Who provides educational materials for geography on the Web?

Web sites exist which cover most topics in geography and related subjects, and many of them offer educational materials for use in schools and higher education. However, not all Web information is from traditional educational sources. An increasing number of information providers are entering the game, including:

- *Educational organisations*

  There are numerous Web sites set up by academics which provide invaluable educational resources. The StatLib site [SL], for example, provides free access to statistical software, data and related information. An increasing number of geography lecturers are putting course material onto their university Web server. Some do it to avoid printing large numbers of handouts, others do it to ensure a single up-to-date source of information for all of their students, and yet others do it to make their work known to a wider community of learners. You will find a rich diversity of educational materials on the Web, including: lecture follow-up notes, supplementary learning materials, diagrams and pictures, and datasets for practical exercises. However, by far the most common is the module or course description document. There is not enough space here to list even a fraction of the Web sites and documents that may be useful for geography courses. A valuable starting point is the home page of the Computer Teaching Initiative (CTI) Centre for Geography, Geology and Meteorology based at Leicester University [CTIG], which provides links to many educational providers.

- *Government organisations*

  Most national statistical organisations provide information for general consumption on their Web sites. In terms of geographical relevance, several federally-funded US organisations (for example, the Bureau of the Census at [USCB], and NASA [NASAI] provide valuable spatial information. Some are even providing explicitly educational products (for example, NASA's KidSat site [NASAKS]). The USGS has established 'The Learning Web' [ULW], which is dedicated to K-12 education, exploration, and life-long learning, and is meant as a gateway to educationally useful earth science information. Global information is provided by many UN organisations; the Food & Agriculture Organisation (FAO) Web site [FAO], for example, provides access to the FAOSTAT database as well as to reports and other publications.

- *Non-governmental organisations (NGOs) and environmental pressure groups*

A number of groups active in the environmental and conservation movements have Web sites, and these can provide case study material in the form of reports of recent activities. Some also produce educational materials. The Rainforest Action Network has a Web site [RAN]; OneWorld Online is a British charity group concerned with global justice and development issues [OWO]; Envirolink [EVL] which is a worldwide environmental information clearinghouse, and connects hundreds of organisations and volunteers; and Friends of the Earth, which has a particularly valuable Web site for those teaching about environmental activism [FOE], though its popularity often leads to slow response times. Mention should also be made of Web sites which have been set up by umbrella organisations concerned with environmental and social issues. An example is EcoNet [ECN], which hosts groups with a concern for ecological sustainability and environmental justice.

- *Commercial organizations*

A number of commercial organizations who specialise in environmental or geographical matters are beginning to offer services to the educational sector, some of which are being charged for. The National Geographic Society [NGS] offers a number of classroom products for geography education in the 'Resources' section of its Web site, as well as a free 'Map Machine'. The BBC has recently launched The Learning Station [TLS], and BT has established a strong presence with its CampusWorld site [BTCW]. Students can also research the environmental aspirations and performance of corporations by visiting their Web site [BT]. The Web is already showing signs of being a growth area for companies selling into the educational market. Examples include EduMall, Magellan University, SyllabusWeb, SyllabusMart and many others. Pressure placed by governments on the educational sector to develop partnerships with commercial organisations in order to reap the benefits of educational technology (as in the case of the UK's National Grid for Learning proposals) increase the need for vigilance to ensure that academic standards are not compromised, and that the mission of subject associations is not compromised, by this trend.

- *Interested individuals*

Search engines often reveal Web pages from sites created by individuals who have developed a personal interest in geographical subjects or environmental issues. Many of these are keen to communicate with others on topics of mutual interest, and many are activists or lay scholars. Examples of such sites are the GeoNet [GN] and the geography site created by a teacher in an Australian school

[SPGSPS]. Eventually, the Web could be populated by as many contributors from the 'informal' information sector as from the 'formal' information economy. It might be worthwhile recommending such sources to students, or even contacting them to dig out further information from their part of the world that could provide invaluable case studies.

It is important to bear in mind that although each type of organisation mounts supposedly 'instructional' materials on its Web sites, many of these are of dubious educational value. Even the materials from higher educational institutions are often little more than terse lecture notes or aides-memoires to students who have missed a particular lecture, and will be relatively meaningless to visitors elsewhere. High quality tutorial materials are few and far between, though the occasional illustrated case study from an environmental organisation or research group may provide useful material for students, and is always worth looking out for.

3.4.4   How should I look for educational materials on the Web?

The Web is rapidly taking over from ftp as a mainstream data provider on the Internet. There is no room in this Guide to list even a fraction of the Web sites of relevance to geography teaching. However, there are three useful ways of tracking down educational resources on the Web:

- *Information Gateways and Directory Services*

  Information gateways (or 'jump off' stations) are Web sites that provide lists of valuable resources on the Web in a particular subject area and are compiled by an individual or organisation to help new users get started. An example of a general gateway for academics in the UK is provided by National Information Services and Systems (NISS), whose 'Information Gateway' was launched in 1995 on its Web site [NISS]. This provides access to a wide range of general information sources for use in higher education, including: library catalogues, databases and electronic journals, HEFCE Quality Assessment Reports, Web sites and mailing lists. The Social Science Information Gateway (SOSIG) [SOSIG] provides links to some materials that may be useful to human geographers. A more specific geography gateway is provided at the Web site managed by the CTI Centre for Geography, Geology and Meteorology at the University of Leicester [CTIG] (Unwin & Maguire, 1990). This has established itself over the past few years as the premier index site for educational materials in the discipline. A useful site for those involved in environmental education is provided by EE-Link [EEL].

  Directory services (also known as catalogue services or search trees) are usually commercial sites which provided a hierarchically structured index to Web resources on a wide range of (usually the most popular) topics. Among the most well-known are Yahoo, Magellan and Excite, though several search engines (for example, Infoseek — see below) are now also providing a directory service. Geography can be located in most of these directories, though its location in the tree-like index structure is not always obvious. (Geography can be found on the Excite site at

[EXG]; with Magellan it is at [MAGG]; and with Yahoo it is at [YAG].) An example of a specialist directory of interest to geographers is the Environmental Organization WebDirectory [EOWD].

- *Search engines*

  A search engine is a piece of software made available on a computer linked to the Internet that will look for information stored in documents on the Web on a particular subject. Among the better-known search engines currently available on the Internet are: Infoseek, Alta Vista, HOTBOT, Lycos and Webcrawler, all of which attempt to index Web information of all kinds. More specialist search engines limit themselves to particular areas of interest. The good news about these engines is that you can use them free of charge (for the moment), though you have to put up with annoying ads (or 'Intermercials') that appear on their search result pages. You can probably gain access to a search engine simply by clicking on the 'search' (or similar) button on your Web browser. The engine then expects you to enter keywords, just like a library catalogue system. You should avoid using search terms which are too general or too specific, and be prepared to wade through a lot of useless Web sites when the engine returns the list of documents that match your query. Because each search engine has its own strengths and weaknesses, it is a good idea to use more than one of them to find the information you are looking for.

- *Surfing*

  The most flexible (and most time-consuming) way of finding useful Web sites is to start with a known Web document that has embedded links (or URLs), then click on a link to move to another relevant document on the Web, and so on. (This is not too dissimilar to following a trail of references in paper books and journals, except that it takes far less time to locate the next reference in the chain.) Useful URLs also appear regularly in newspapers, journals, textbooks, magazines, and online sources, and it is relatively easy to compile a personalised list of such links for future reference by 'bookmarking' them on your Web browser.

3.4.5 Should I convert my course materials for student use on the Web?

"We are not yet convinced that there is any pedagogical benefit arising from simply hypertextualizing the existing course material."

(Benyon et al., 1997, p.214)

A growing number of geographers and geography departments are transferring course-related materials onto their institution's Web server. (The Virtual Geography Department project based at the University of Texas is a major example of this trend.) In practical terms, this is extremely simply to do. A growing number of word processors permit a document to be converted from native format to HTML format and vice versa at the click of an icon, and easy-to-use HTML editors and GIF creation kits are also readily available. It also requires less training and time to create course materials as Web documents than to create polished multimedia materials using commercial authoring software. This said, however, eye-catching and effective Web pages are not easily produced, and the task of creating them is therefore perhaps best left to someone with considerable document design sense.

For some course materials (for example, course handouts, reading lists, timetabling and other administrative arrangements), there are a number of clear advantages to placing them on the Web:

- continuous availability — students can keep in touch with their course everywhere on campus, while studying off campus, and at times when the institution is shut;

- easy update — the content of Web pages can be easily corrected and extended;

- rapid dissemination — news can be posted immediately, without the delay of going through conventional reprographics;

- external publicity — people in other parts of the country/world can see what you are up to, and maybe enrol on one or more of your courses as a result;

- environmental responsibility — using an electronic medium reduces the amount of paper used to support course delivery. However, students often end up printing what they find on the Web, which negates this advantage, and costs more overall than if the printing were organised communally.

You can choose between the World Wide Web or a local Web as a home for course-related materials. A local Web is useful for registering students, for providing student feedback and for posting assessed work. The World Wide Web, by contrast, is a better place for course information and handouts, reference study materials, learning resources and past exam papers. You should base your decision as to what goes where according to the following criteria: quality, privacy, cost, confidentiality, libel and copyright.

The temptation to go further than course handouts, and to transfer teaching materials onto the Web, needs more serious thought. If individual lecture handouts or even a course reader can be given added value by being turned into Web documents (for example, by adding hypertext links or including animated illustrations), then the outcome will probably be worthwhile. However, if the end result is simply an 'electronic book', then the effort will probably be wasted. The computer screen is not a convenient reading medium for most students, and it is not as easy for students to annotate what appears on screen as it is with paper. A useful review of one lecturer's experiments can be found at [BOTH]

Committing student-centred learning materials to the Web is another issue altogether. Even when such materials have been carefully designed for independent study, it is not always clear whether their conversion to Web form will deliver expected benefits. This, in fact, was the judgement of a course team at the Open University (Benyon *et al.*, 1997) who experimented with converting existing learning materials into Web format, and found that current Web technology is wanting in the following ways:

- Little control over the format of material appearing on a student's screen;

- Inability to use various types of links, or to use dynamically variable links;

- Poor range of interaction devices (for example, pop-up menus).

A more limited approach may therefore be preferable in the short term, until Web technology matures further. Here are three examples of what you might consider doing:

- Place your conventional reading lists on the Web, together with the text of references for which you have obtained copyright clearance. (You could even create an electronic version of a course reader.) One experiment currently under way (the ACORN Project at Loughborough University [AP]) is exploring the potential of loading onto the Web geography reading lists and the text of journal papers they refer to (Kingston, 1997).

- Place personal support materials for particular lectures or other classes on the departmental Web site. For example, you could scan photographs taken in various parts of the world you have visited, and place them in a Web gallery. Examples are provided by the tutors on a course on the Architecture of Russia, the Ukraine and Uzbekistan [AR], and by the Geo-Images Project [GIP]. Another useful idea is to produce audio-visual aids for use in lectures as Web documents rather than using conventional presentation software (such as *PowerPoint*), because students can then visit your Web site after the class to review the material at leisure. See, however, Webb (1997) for an example of the use of presentation software to create lecture material that is subsequently reviewed by students on a local fileserver.

- Maintain a list of Web sites that are relevant to your courses, and put it on your institution's Web site for general student access. An increasing number of lecturers are now doing this, particularly in North American universities. Two UK examples are the list for Environmental Science students maintained by Ian Livingstone and colleagues at Nene College [NCES], and the list created for geography students at Lancaster University as part of a TLTP project to develop an introductory Internet tutorial [LWT].

Case Study 8

Collaboration among staff to produce a Web fieldcourse site

John Stainfield, in the Department of Geography at Plymouth University, has begun a Web project to create a collection of shared resources for fieldwork. Two features set this initiative apart. First, it focuses on a single fieldcourse venue, Malta, which is regularly visited by Plymouth students. John has already found at least 10 other departments of geography and environmental science in the UK that take similar trips to the island. This brings us to the second feature of this project: the collaboration of as many people as possible who are involved in such trips to create a joint resource that all can use to their mutual advantage.

The current idea is to bring together two kinds of resource on a Web site. The first will be descriptive materials that introduce students to the geography of the island, and to its emerging social and environmental problems. The second will include interactive exercises (for example, a vegetation quadrat sampling simulation) to enable students to understand more of the island through active

(cont.)

learning. The former material will probably consist mainly of standard Web documents (in HTML format); the latter will probably consist of small downloadable Java applets or executables.

Eventually, it is hoped to extend the Web site to include material relating to elements that Malta shares in common with other parts of the Mediterranean, such as tourism, water pollution, resource exploitation, and soil erosion.

The key idea behind this venture is staff collaboration. So, if you would like to join in the project, contact John now!

Innovator: John Stainfield, Department of Geography, University of Plymouth.
Email: jstainfield@plymouth.ac.uk

3.4.6   How can my students use the Web?

The Web is typically viewed by lecturers and students as a source of educationally valuable information. However, it is equally important as a shop-window for materials produced by students as part of their learning activities. In some universities, students are required to 'publish' their assessed coursework on the Web, or even to set up a mini-Web site as a project assignment. The former is sometimes justified on the grounds that it increases the scope for students receiving feedback from peers and experts, and the latter is justified on the basis that the best way of understanding a subject is to 'teach' it to others. The first of these can be undertaken through a department's Webmaster, the second usually requires that an institution or department provides space where students can set up their own home page, and requires students to learn how to create Web documents.

Posting coursework on the Web can be both challenging and stimulating for students, and has the practical benefit of allowing easy staff access to coursework from both remotely located and campus-based students. Opinion is divided over whether student work should be 'published' on the Web, for all to see, or whether it should be restricted to an internal Web. At the postgraduate level the issue is not so contentious; many Masters and doctoral dissertations are now being posted on the Web, and this provides valuable review material for instructors, particularly on newly emerging geographical issues, such as the social implications of on-line communities. However, the notion of students training as Webmasters is perhaps more contentious, because of the time needed for them to become proficient in setting up a personal Web site.

In the Department of Geography and Earth Systems Science at George Mason University several staff encourage their students to place completed course projects on the Web. Recent examples include electronic regional essays from around the world for Lee De Cola's course on GIS Using Internet Resources, and final project maps from Jeremy Crampton's course in Advanced Cartography [GMU].

Case Study 9

Creating electronic posters on the Web

At Oxford Brookes University, Roger Suthren gets his students to produce posters on the Web rather than on paper. The disadvantages of paper posters are well known:

- material costs
- limited student access
- storage space
- damage through repeated use
- difficult to update

Electronic posters on the Web have several counter-balancing advantages:

+ always accessible (but see below)
+ easy to update
+ can be interactive
+ easy to copy and print

They are not, however, without disadvantages of their own:

- prior training required in using Web browser and editor
- screen size limitations
- problems of student access to computers
- difficulty for student groups to design and view collectively

Innovator: Roger Suthren, School of Construction & Earth Sciences, Oxford Brookes University. Email: rjsuthren@brookes.ac.uk

3.4.7 What problems will my students face if I let them loose on the Web?

Time wasting

It is well known that surfing the Internet and/or the Web can be both incredibly beguiling and immensely wasteful of student time. As tutors, we must take on some of the responsibility for ensuring that the assignments we set involving use of the Web do not lead students to spend more time than they should in tracking down appropriate information. Various strategies can be adopted. One is simply to set clear time limits on the amount of time that should be spent on a Web-related assignment. Another, adopted by Chad Staddon, a geographer at the University of West England (C-STADDON@wpg.uwe.ac.uk), is to provide students with informal workshops on how to optimise their time on the Web. At the Universities of Plymouth, Aberystwyth and Cambridge, a group of geographers provide more formal training on Web use, but with off-line rather than live sessions (Newnham *et al.*, 1998).

Inflated expectations

Many students' expectations of the Web are far too high. Tutors should impress on them that they should not expect to find automatic answers on the Web to their coursework assignments or projects, and that the vast majority of material on the Web is almost worthless. They should also encourage students to operate with the prime principle of investigative journalism: ensure that any information comes from at least two independent sources — and that this does not mean from a Web site and one of its mirror sites!

The medium or source from which students acquire resource materials can significantly affect their perception of its usefulness, and this can have a negative impact on its educational contribution. Of the three means of delivery in common use, CD-ROM, Internet/ Web and LAN, the first is currently far and away the most successful, because it ensures rapid and consistent response times. At the time of writing, the Internet is not very effective at delivering multimedia materials that include audio and video. Although RealAudio provides reasonable sound over the Internet, and software such as CUSEEME from Cornell University permits video-conferencing over the Internet, it will be some time before time-varying information will be effectively accessed from the Internet.

Plagiarism and Personation

Over the past decade, the proportion of student assessment in geography involving coursework has risen dramatically. When setting assessed coursework assignments, and particularly the traditional undergraduate dissertation, tutors have always faced the problem of students submitting the work of others (for example, dissertations lodged in other university libraries) as their own. Easy access to information on the Web, and particularly to review essays and postgraduate theses mounted on Web sites by academics, provides increasing opportunity for students to pass off other people's work as their own. A related and growing problem is that an increasing number of sites on the Web offer ready-written essays or project papers on a growing number of subjects [RA]. Some sites offer papers written by past students freely, others charge for supplying papers, and yet others will write new papers for students for a suitable fee. Although these forms of cheating are not new, the Internet provides far easier access to materials for weak or lazy students. Tutors can adopt the following strategies to help circumvent this problem:

- Tailor assignments closely to work undertaken on a particular course.

- Change assignments regularly, even in a small way.

- Monitor student progress during the preparation of submitted coursework.

- Compare the writing and thinking styles of suspicious materials with a student's past coursework.

- Familiarise yourself with some of the Web sites offering students access to papers in geography and related disciplines (ask the library staff for help in doing this).

- If you suspect that a submitted item is derived from Web material, undertake a Web search (with one or more search engines) using some of the key terms in the coursework, and see if the suspect material turns up.

Regurgitation

Another problem is the 'cut-and-paste' mentality of students who find relevant material on the Web, and submit it without any interpretation of their own. Admittedly, this is not a problem that lies solely at the door of the Web (students often make similarly mindless use of CD-ROMs in the university library), but it is one that is encouraged by easy access to Web material. The solution lies with the educational process itself, and with the demands that tutors place on their students. At the very least, all students should be encouraged, from their first year, to critique all information that comes their way, and to ensure that healthy scepticism permeates all their submitted coursework. This will ensure that dubious quality Web material is weeded out at an early stage, and that material which **is** used is properly integrated into the student's own line of thought.

Inadequate referencing

When students submit material from the Web, they rarely provide proper referencing to their sources. Again, this is a common problem with conventional sources of information, but the problem is perhaps heightened with the Web because many students (and staff) are uncertain as to how Web materials should be referenced. The simplest advice is that all Web sources should be referenced in the text just like any other material, and that the reference list should contain the full URL of the source document and, where available, its author and source date. It is also advisable for students to include the date they accessed the material, because Web sites are notoriously fickle; contents change through time, and pages and sites are also prone to relocating or disappearing without warning. The problem of not being able to refer to specific 'pages' in a Web document is still largely unresolved.

3.4.8   What are the key educational issues related to student use of the Web?

Maddux (1994) provides a general discussion of a range of operational problems associated with educational use of the Internet in its current form. In this section, a number of more fundamental educational issues are addressed.

Getting lost in hyperspace

The first problem is that many students (and their tutors) get lost in the infinitely complex network of linkages between documents on the Web, rapidly losing track of their current location, failing to grasp an overall context of the information they are searching, and often ending up moving aimlessly from document to document. This happens particularly with those who are not used to independent learning, or to controlling their personal search space. As a solution to this problem, several educational resource developers (for example, Hammond & Allinson, 1987) have suggested that the Web's information complexity should be curbed by providing students with some or all of the following facilities:

- structured interfaces to the Web and other hypermedia
- indexes to information in Web documents
- guided tours to hypermedia-linked documents
- information maps

Others have suggested restricting students (initially at least) to a small subset of Web information, which can be stored locally on a departmental or institutional computer (for example, Newnham *et al.*, 1998).

The lure of intervening opportunities

When students begin to search for material on the Web, even within the context of a clearly limited assignment, they invariably get sidetracked by the appeal of links to documents on related or unrelated topics. Of course, this is one of the strengths of the Web, but to students new to the this environment it can be an almost fatal attraction. The mental stimulation or buzz that comes from randomly following links is something that is beginning to attract serious academic study. But in the context of a geography assignment, it can prevent students from undertaking focused study and developing clear lines of thought.

Browsing versus learning

There is a widespread assumption in the educational literature that 'browsing' through information (especially the hypermedia information commonly found on the Web) automatically leads to effective learning. Although there is little research evidence as yet, there are growing suspicions that the much-vaunted browsing paradigm of learning through hypermedia (on CDs or on the Web) is deeply flawed. Why, it is asked, should students be expected to learn anything simply by browsing through a network of information on a particular topic? Is learning in this context any more likely to occur than from viewing an assortment of film or video clips, or from leafing through a collection of randomly accumulated paper cuttings in a box file placed in the departmental resource centre? At the heart of the debate over browsing and learning lies a fundamental belief about the nature of information and the nature of human learning.

Many believe that the Web (like hypermedia in general) provides information in a non-sequential form which reflects the semantic network that humans are supposed to carry around in their heads. By providing students with information that mimics a semantic network, it is held that the brain will more readily assimilate such information. Two objections can be made to this assumption. The first is that the information 'chunks' and network connections in hypermedia and the Web are not necessarily the same as in the human brain. There is still no satisfactory definition of the differences between 'data', 'information' and 'knowledge', and precisely which of these (if any) is stored in the brain. Nor do we have anything more than a crude understanding of the supposed structure of 'semantic networks' in the human brain. The nature of the links between chunks of information in hypermedia and on the Web are more transparent, because developers have written about them in detail. Not only are there many types of such links, but they can vary from document to document and from courseware to courseware, and few hypermedia developers appear to model their links on known characteristics of information linkage in the human brain.

The second objection to the benefits of information networks is that even if the brain develops its own form of semantic network, it does not necessarily follow that by presenting it with information in network form (assuming that it is the 'right' kind of network) that it will more readily assimilate that information. Research into learner differences (for example, Pask, 1976; 1988) suggests the existence of serial and holistic learners; evidence for 'network learners' is still awaited. Further development of this critique is beyond the scope of this Guide; further discussion can be found in McKendree *et al.* (1995). A practical implication is that you should consider providing students with activities that encourage them to use the information available on the Web — for example, provide them with opportunities for active learning rather than passive and aimless browsing.

Independence in learning

Another belief underlying the 'browsing equals learning' model of Web-based education is the role of the student as autonomous learner. It has become something of an article of faith in recent years that independent learning is a 'good thing', and that the development of the 'autonomous learner' should be a primary goal of higher education. This belief system engages naturally with the Web, which appears to provide a natural context for students to develop the skills of independent learning, and to express their autonomy by exploring information in ways not prescribed by their courses or their tutors. Because anyone can post material on the Web, it is an environment which removes the distinction between 'learner' and 'teacher'. It appears to provide a democratisation of the learning process, where the student can be both learner and teacher. This is far removed from the conventional classroom where power relations greatly favour the teacher.

The notion that independent learning necessarily implies learning alone, and that autonomous learning must automatically dispense with any form of external control, are surely misguided. The Web is a resource like any other, and should be used with both care and control. Granting complete freedom to students, whether in relation to the Web or to any other educational resource, is surely an abdication of our real responsibility as teachers, which is to nurture and to guide.

Quality control

It has frequently been stated that the value of information on the Web is inversely proportional to its quantity. The information available on the Web is problematic for many students, particularly those who have not yet acquired a sufficient grounding in a particular area of study to enable them to discriminate between 'good' and 'bad' information, or between 'useful' and 'useless' information. Shepherd & Castleford (1998) discuss the problems involved in defining quality information in relation to the Web.

In using the Web, most students will benefit from the guidance and contributions of others, whether they be their peers or their tutors. If students are to focus on the interpretation and application of ideas and information found on the Web, then it is important that they learn how to discriminate between poor and good quality sources. You can do this in several ways; two of the most common approaches are:

- Shield your students from poor quality information by selecting relevant information on their behalf. This approach has been taken by Newnham *et al.* (1998), who downloaded a selection of materials on volcanoes and earthquakes onto an institutional fileserver to which students had network access. The same selective approach is taken by the Web directories and information gateways discussed above (see Section 3.4.4).

- Provide students with effective information selection and evaluation skills, either as part of a study skills course, or built into the substantive geography modules in which they use the information (see Whalley, 1997). (Ball, 1994, discusses the pros and cons of using students to gather information from media sources.)

Other strategies for steering students towards quality information on the Web are discussed in Shepherd & Castleford (1998), who also consider the training necessary for students to become information literate. Beyond this lies the broader issue of what Ò Tuathail & McCormack (1998a; b) calls 'technological literacy': the ability of students to look beyond the value-free appearance of information on the Web to understand the role of global capitalism in shaping the new information landscape for its own ends.

3.4.9 Should we be aiming to establish a 'Virtual' Geography Department?

"If learning is to become a highly differentiated, anywhere, anytime activity it will be necessary to reengineer more than just the syllabus, delivery mode, and teaching model. We will need to make the digital or virtual library a reality as well."

(Heterick, 1993)

In a growing number of institutions around the world, experiments are under way in a wide range of disciplines to set up 'virtual classrooms', 'virtual courses', 'virtual departments' and even 'virtual universities' (for example, Hiltz, 1994; Tiffin & Rajasingham, 1995; and the Virtual Online University [VOU]). The common element in all of these initiatives is the transfer of learning resources and/or learning activities onto a communications network, usually the World Wide Web. (The other common element is the unfortunate use of the 'V' word.)

One of the more achievable aims is the provision of a single electronic location where students enrolled on a department's courses can find all of the relevant information for their studies: enrolment details, course synopses, timetables, class handouts, examination regulations and past papers, etc. This not only goes some way to ensuring that no group of students is penalised due to their chosen mode, location or pattern of study, but it also helps to reduce the departmental output of paper.

A more challenging objective is the delivery of individual modules or entire courses on the Web, as discussed earlier. The geography department at the University of Texas is coordinating a number of departments in the US and overseas who aim to provide curriculum materials over the Web, and its Web site is a valuable place for monitoring this experiment [VGD]. To date, most of these developments fall well short of a completely 'virtual' (i.e. exclusively electronic) department.

The ultimate goal of the 'virtual department' is probably the provision of an opportunity for people to acquire higher education entirely through the Internet. In some cases, this would inevitably lead to the replacement of fixed-location departments on a particular campus either with a dispersed department, connected by telecommunications, or to a department with no physical location at all. The first of these approaches is in line with the model of the 'virtual corporation' which has been advocated recently in some management circles; the second approach is reflected in some of the experimental learning networks (for example, Usenet University, the Globewide Network Academy, and Diversity University) that have been developed recently in the USA, and which are entirely new creations, having had no prior campus existence. Gibbs (1997) provides a valuable example of how departments might set about deciding if they want to go down this route.

Perhaps a more balanced approach is represented by those geography departments who use the Internet and the Web to enhance learning opportunities for geographically dispersed students, rather than attempting to transfer all learning experiences onto the Web. Examples of recent experiments include the use of the Web in association with a peripatetic teacher to run a module on the regional geography of Africa at George Mason University (Crampton, 1998), and the trans-Atlantic course unit on Third World representations on the Web which was mediated entirely by communication technologies (Steeples *et al.*, 1998).

Summing up

"Computing is no longer a tool... Today computing is an environment."

(Ritter, 1996)

"Learning is not a spectator sport."

(Heterick, 1993)

"Somehow we have to find a way to create this new synthesis, in which all modes of learning are honored and given an appropriate place."

(Monke, 1996)

By way of a conclusion, we offer a few general principles about the educational use of ICT in geography which are drawn from the material presented in various sections of this Guide.

1.   Computers have been at the forefront of technology-based education over the past 30 years. The torch is now being passed to communication technologies, which will arguably have a far greater and more beneficial impact on teaching and learning.

2.   Used carefully, information and communication technologies can be equally as effective as other approaches in fostering student learning, but there is only limited evidence to suggest that they will out-perform other methods.

3.   Information and communication technologies are used to best effect as part of a general approach to teaching that is developed independently of any particular set of tools, and which integrates various teaching and learning methods (for example, see Rich *et al.*, 1997). The tools available should be chosen on the basis of how well they fit one's educational philosophy, students' preferred learning styles and the subject matter being studied.

4.   There are many opportunities, ample resources, and a growing number of examples of best practice to encourage and guide the adoption of ICT in geography. However, care will need to be taken to evaluate the increasing amount of 'educational' products and services that are being provided by commercial organisations.

5.   Technology should be treated as an educational tool rather than as an educational philosophy in its own right.

Finally, it is perhaps worth listing the five characteristics of effective technology-based courses identified by Jordan & Follman (1993), because they highlight the beliefs that underlie most of what has been said or implied throughout this Guide. (A similar list of seven principles of good practice in undergraduate education has been published by the AAHE [TSP].) Good technology programmes:

- emphasise cooperative learning models, allowing heterogeneous groups of students to work together collaboratively;

- emphasise higher-level problem-solving skills while at the same time reinforce basic skills;

- support interactions between students and teachers rather than use computers as teaching machines to replace teachers;

- create interactive learning environments built around real-world problems;

- adapt to a variety of student learning styles.

Of course, none of these are specific to technology. All approaches to teaching and learning share a common ground, and all need to be used in ways that are consistent with a fundamental philosophy of what makes for effective student learning.

5 References

5.1   References: books and journals

5.1.1   References cited in the text

Agnew, C. & Elton, L. (1997) *Lecturing in Geography* (Cheltenham: Geography Discipline Network, CGCHE).

Ambron, S. & Hooper, C. (1988) *Interactive multi-media* (Redmond, WA: Microsoft Press).

Bailey, E.K. & Cotlar, M. (1994) Teaching via the Internet, *Communication Education*, 43, pp.184-193.

Ball, R. (1994) Monitoring the media: student production and use of newspaper resource files in Economic Geography, *Journal of Geography in Higher Education*, 18(2), pp.222-228.

Bates, A.W. (1994) Educational multi-media in a networked society, in: *Proceedings of ED-MEDIA 94*, pp.3-8 (Charlottesville, VA: Association for the Advancement of Computing in Education).

Batty, M., Bracket, I., Guy, C. & Spooner, R. (1985) Teaching spatial modelling using interacting computers and interactive computer graphics, *Journal of Geography in Higher Education*, 9(1), pp.25-36.

Beattie, K.C., McNaught, S. & Wills, S. (Eds) (1994) *Interactive Multimedia in University Education: designing for change in teaching and learning* (The Netherlands: Elsevier Science).

Benyon, D., Stone, D. & Woodroffe, M. (1997) Experience with developing multimedia courseware for the World Wide Web: the need for better tools and clear pedagogy, *International Journal of Human-Computer Studies*, 47(1), pp.197-218.

Berners-Lee, T. (1996) WWW past, present and future, *Communications of the Association of Computing Machinery*, 29(10), pp.69-77.

Birnie, J. & Mason O'Connor, K. (1998) *Practicals and Laboratory Work in Geography* (Cheltenham: Geography Discipline Network, CGCHE).

Bishop, M.P., Shroder, J.F. & Moore, T.K. (1995) Integration of computer technology and interactive learning in geographic education, *Journal of Geography in Higher Education*, 19(1), pp.97-110.

Boettcher, J.V. (Ed.) (1992) *101 Success Stories of Information Technology in Higher Education* (New York, NY: McGraw Hill).

Boyd, E.M. & Fales, A.W. (1983) Reflective learning: key to learning from experience, *Journal of Humanistic Psychology*, 23(2), pp.99-117.

Bradbeer, J. (1996) Problem-based learning and fieldwork: issues of representation in teaching race and ethnicity, *Journal of Geography in Higher Education*, 20(1), pp.1-18.

Bradford, M. & O'Connell, C.. (1998) *Assessment in Geography* (Cheltenham: Geography Discipline Network, CGCHE).

Bradshaw, M. (1995) Common core or food for the political worm?, *Journal of Geography in Higher Education*, 19(1), pp.5-9.

Briggs-Myers, I. & Myers, P.B. (1980) *Gifts Differing* (Palo Alto: Consulting Psychologists Press).

Browne, T. (1992) The great divide: central versus departmental support for teaching GIS, *Journal of Geography in Higher Education*, 16(2), pp.225-231.

Chalkley, B. & Harwood, J. (1998) *Transferable Skills and Work-based Learning in Geography* (Cheltenham: Geography Discipline Network, CGCHE).

Charman, D. (1997) *Improving student learning with computer-based feedback*, http:// www.chelt.ac.uk/gdn/abstracts/a47.htm.

Chapman, K. (1997) *Changing assessment practices in first and second year geography modules using Computer-Assisted Assessment (CAA) to set formative and summative objective tests*, http://www.chelt.ac.uk/gdn/abstracts/a40.htm.

Clark, G. & Wareham, T. (1997) *Small-group Teaching in Geography* (Cheltenham: Geography Discipline Network, CGCHE).

Clinton, W. (1997) *Technological Literacy, President Clinton's Call for Action for American Education in the 21st Century* (http://www.ed.gov/updates/ PresEDPlan/part11.html).

Conway, D. (1994) Student response to hypermedia in the lecture theatre: a case study, *Proceedings of ED-Media*, 94, pp.141-146.

Cook, K. (1997) *Teaching development issues using information technology*, http:// www.chelt.ac.uk/gdn/abstracts/a23.htm.

Coventry, L. (1997) Video conferencing in higher education, *SIMA Report* (http:// www.man.ac.uk/MVC/SIMA/video3/contents.html).

Crampton, J. (1998) A regional geography class in a distributed learning environment, *Journal of Geography in Higher Education*, 22(3), in press.

Cuban, L. (1986) *Teachers and Machines: the classroom use of technology since 1920* (New York; London: Teachers College Press).

Davies, M.L. & Crowther, E.A. (1995) The benefits of using multimedia in higher education: myths and realities, *Active Learning*, 3, pp.3-6.

DfEE (1997) *National Grid for Learning* (London: HMSO). Also available on the Web at: http://www.open.gov.uk/dfee/grid/what.htm

Donert, K. (1998) *A Geographer's Guide to the Internet* (Sheffield: Geographical Association).

Drummond, I. (1997) *Computerised collections and alanysis of teaching quality assessment data*, http://www.chelt.ac.uk/gdn/abstracts/a43.htm.

Elsom-Cook, M. (1989) Guided discovery tutoring and bounded user modelling, in: D. Laurillard (Ed.) *Interactive Media*, pp.165-178.

Entwistle, N. & Ramsden, P. (1983) *Understanding Student Learning* (Croom Helm).

Felder, R.M. (1996) Matters of style, *ASEE Prism*, 6(4), pp.18-23. Also available on the Web at: http://www2.ncsu.edu/unity/lockers/users/f/felder/public/Papers/LS-Prism.htm.

Ferrigno, C.F. & Wiltshire, D.A. (1994) Development and evaluation of a series of hypermedia educational systems for the Earth Sciences, in: *Proceedings of ED-MEDIA 94*, pp.203-208 (Charlottesville, VA: Association for the Advancement of Computing in Education).

Flowerdew, R. & Lovett, A. (1992) Some applications of Murphy's Law: using computers for geography practical teaching, *Journal of Geography in Higher Education*, 16(1), pp.37-44.

Foote, K.E. (1994) The Geographer's craft, *The Pennsylvania Geographer*, 32, pp.1-25. Also available on the Web at: http://www.utexas.edu/depts/grg/gcraft/about/penngeog/penngeog.html.

Gardiner, V. & Unwin, D.J. (1986) Computers and the field class, *Journal of Geography in Higher Education*, 10(2), pp.169-179.

Gibbs, G. (1997) The role of IT at Brookes in 2001 — an alternative scenario, *Teaching Forum*, 44, Spring, pp.11-13.

Gibbs, G. & Jenkins, A. (1984) Break up your lectures: or Christaller sliced up, *Journal of Geography in Higher Education*, 8(1), pp.27-39.

Gilder, G. (1992) *Life after Television* (New York: Norton).

Gold, J.R., Jenkins, A., Ley, R., Riley, J., Shepherd, I. D. H., & Unwin, D. J. (1991) *Teaching Geography in Higher Education: a manual of good practice* (Oxford: Blackwell) — particularly Chapter 7: Computer-assisted Learning and Teaching.

Greenberg, J.M. (1994) Integrated multimedia in distance education, in: *Proceedings of ED-MEDIA 94*, pp.21-25 (Charlottesville, VA: Association for the Advancement of Computing in Education).

Griffith, D.A. (1992) Teaching spatial statistics to geographers using MINITAB, *Journal of Geography in Higher Education*, 16(1), pp.45-60.

Hall, B.W. (1993) Using E-mail to enhance class participation, *PS: Political Science and Politics*, 26(4), pp.757-761.

Hammond, N. (1989) Hypermedia and learning: who guides whom? in: H. Maurer (Ed.) *Computer assisted learning* (Berlin: Springer-Verlag).

Hammond, N. & Allinson, L. (1987) The travel metaphor as design principle and training aid for navigating around complex systems, in: Diaper, D. and Winder, R. (Eds) *People and computers III*, pp.75-90 (Cambridge: Cambridge University Press).

Hammond, N. & Allinson, L. (1989) Extending hypertext for learning: an investigation of access and guidance tools, in: A. Sutcliffe & L. Macaulay (Eds) *People and Computers V*, pp.293-304 (Cambridge: Cambridge University Press).

Hardisty, J., Taylor, D.M. & Metcalfe, S.E. (1993) *Computerised Environmental Modelling*, (Chichester: Wiley).

Hay, I. & Delaney, E.J. (1994) "Who Teaches, Learns": writing groups in geographical education, *Journal of Geography in Higher Education*, 18(3), pp.317-334.

Healey, M. (1998) *Resource-based Learning in Geography* (Cheltenham: Geography Discipline Network, CGCHE).

Heard, S., Nicol, J. & Heath, S. (1997) *Setting Effective Objective Tests* (University of Aberdeen: Centre for Educational Development, CTI CLUES, and Department of Geography).

Heath, J. (1995) When interactive multimedia is not truly interactive, *Active Learning*, 3, pp.7-9.

Henning, E. (1997) Schools online?, *PC Magazine*, 6(9), p.89.

Heterick, R.C. Jr. (1993) Introduction: Reengineering Teaching and Learning, in: R. C. Heterick (Ed.) *Reengineering Teaching and Learning in Higher Education: Sheltered groves, Camelot, windmills, and malls*, CAUSE Professional Paper Series No. 10 (Boulder, CO: CAUSE). (Also available on the Web at: http://cause-www.colorado.edu/information-resources/ir-library/abstracts/pub3010.html)

Hiltz, S.R. (1994) *The Virtual Classroom: learning without limits via computer networks* (Norwood, NJ: Ablex Publishing).

Johnson, D.W. & Johnson, R.T. (1975) *Learning Together and Alone: cooperation, competition, and individualization* (Englewood Cliffs, NJ: Prentice Hall).

Jordan, W.R. & Follman, J.M. (Eds) (1993) *Using Technology to improve teaching and learning. Hot topics: usable research* (Victoria, BC, Canada: British Columbia Ministry of Attorney General). (See also ERIC Abstract No. ED355930.)

Kemp, K.K. & Goodchild, M.F. (1991) Developing a curriculum in geographic information systems: the National Center for Geographic Information and Analysis Core Curriculum Project, *Journal of Geography in Higher Education*, 15(2), pp.123-134.

Kingston, P. (1997) Electronic access to geography course readings: Project ACORN, *GeoCal*, 16, pp.21-23.

Kirkby, M.J., Naden, P.S., Burt, T.P. & Butcher, D.P. (1987) *Computer Simulation in Physical Geography* (Chichester: Wiley).

Kolb, D.A. (1984) *Experiential Learning — experience as the source of learning and development* (Englewood Cliffs, NJ: Prentice Hall).

Kouses, R.T., Myers, J. D. & Wulf, W. A. (1996) Collaboratories: doing science on the Internet, *IEEE Computer*, 22(8), pp.40-46.

Krygier, J.B., Reeves, C., Dibiase, D. & Cupp, J. (1997) Design, implementation and evaluation of multimedia resources for Geography and Earth Science education, *Journal of Geography in Higher Education*, 21(1), pp.17-38.

Kulik, C. & Kulik, J.A. (1987) Review of recent research literature on computer-based instruction, *Contemporary Educational Psychology*, 12, pp.222-230.

Kulik, C. & Kulik, J.A. (1991) Effectiveness of computer-based instruction: an updated analysis, *Computers and Human Behavior*, 7, pp.75-94.

Laurillard, D. (1993) *Rethinking University Teaching: a framework for the effective use of educational technology* (London: Routledge).

Lee, M.P. & Soper, J.B. (1987) Using spreadsheets to teach statistics in geography, *Journal of Geography in Higher Education*, 11(1), pp.27-33.

Livingstone, I. (1997) GeogNet: an email discussion list for UK geographers in higher education, *Journal of Geography in Higher Education*, 21(2), p.291.

Livingstone, I., Matthews, H. & Castley, A. (1998) *Fieldwork and Dissertations in Geography* (Cheltenham: Geography Discipline Network, CGCHE).

Livingstone, I. & Shepherd, I.D.H. (1997) Using the Internet, *Journal of Geography in Higher Education*, 21(3), pp.435-443.

Maddux, C.D. (1994) The Internet: educational prospects — and problems, *Educational Technology*, 34(7), pp.37-42.

Malone, T.W. (1981) Towards a theory of intrinsically motivating instruction, *Cognitive Science*, 4, pp.339-369.

Mayes, J.T. (1992) The 'M' word: multimedia interfaces and their role in interactive learning systems, in: A. D. N. Edwards & S. Holland (Eds) *Multimedia Interface Design in Education*, pp.1-22 (Berlin & Heidelberg: Springer-Verlag).

McCormick, S. & Bratt, P. (1988) Some issues relating to the design and development of an interactive video disc, *Computers in Education*, 12(1), pp.257-260.

McGee, P.A. & Boyd, V. (1995) Computer-mediated communication: facilitating dialogues, in: *Technology and Teacher Education Annual — 1995*, pp.643-647 (Charlottesville, VA: Association for the Advancement of Computing in Education).

McKendree, J., Reader, W. & Hammond, N. (1995) The "homeopathic fallacy" in learning from hypertext, *Interactions*, 2(3), pp.74-82. (Also available on the Web at: http://www.ioe.ac.uk/tescwwr/Homeopathy.html).

Meleis, H. (1996) Toward the information network, *Communications of the Association of Computing Machinery*, 29(10), pp.59-67.

Mellor, A. (1991) Experiential learning through integrated project work: an example from soil science, *Journal of Geography in Higher Education*, 15(2), pp.135-149.

Messing, J. & McLachlan, R. (1994) History, hypermedia, and the birth of a nation, in: *Proceedings of ED-MEDIA 94*, pp.391-396 (Charlottesville, VA: Association for the Advancement of Computing in Education).

Monke, L. (1996) *Computers in education: the Web and the plough* (http://www.public.iastate.edu/~lmonke/online_doc.html#Plow).

Mulligan, M. (1997) *Active student contribution in lectures*, http://www.chelt.ac.uk/gdn/abstracts/a6.htm.

NCIHE (National Committee of Inquiry into Higher Education) (1997) *Report of the National Committee of Inquiry into Higher Education* (London: HMSO). Often referred to as 'The Dearing Report'. (Also available on the Web at: http://www.leeds.ac.uk/ncihe/).

Newnham, R.M., Mather, A.E., Grattan, J.P., Holmes, A. & Gardner, A.R. (1998) An evaluation of the use of Internet sources as a basis for geography coursework, *Journal of Geography in Higher Education*, in press.

Nielsen, J. (1990a) The art of navigating through hypertext, *Communications of the Association of Computing Machinery*, 33(3), pp.296-310.

Nielsen, J. (1990b) *Hypertext and Hypermedia* (San Diego, CA: Academic Press).

O'Day, R. (1997) *Charles Booth and social investigation in nineteenth-century Britain*, Craft, 16, pp.3-8.

Olson, J.M. (1997) Multimedia in geography: good, bad, ugly, or cool? *Annals of the Association of American Geographers*, 87(4), pp.571-578.

Ò Tuathail, G. & McCormack, D. (1998a) Global conflicts on-line: technoliteracy and developing an Internet-based conflict archive, *Journal of Geography*, 97(1), pp.1-11.

Ò Tuathail, G. & McCormack, D. (1998b) The technoliteracy challenge: teaching globalisation using the Internet, *Journal of Geography in Higher Education*, 22(3), in press.

Oppenheimer, T. (1997) The computer delusion, *The Atlantic Monthly*, 280(1), pp.45-62. (Also available on the Web at: http://www.theatlantic.com/97Jnl/computer.htm)

Papert, S. (1980) *Mindstorms* (Brighton: Harvester Press).

Pask, G. (1976) Styles and strategies of learning, *British Journal of Educational Psychology*, 56, pp.128-148.

Pask, G. (1988) Learning strategies, teaching strategies, and conceptual or learning style, in: R. Schmeck (Ed.) *Learning Strategies and Learning Styles* (New York: Plenum Press).

Pea, R.D. (1993) The Collaborative Visualization Project, *Communications of the Association of Computing Machinery*, 36(5), pp.60-63.

Pea, R.D. & Gomez, L.M. (1992) Distributed multimedia learning environments: why and how?, *Interactive Learning Environments*, 2(2), pp.73-109.

Peach, C. (1996) Does Britain have ghettos? *Transactions of the Institute of British Geographers*, 21, pp.216-235.

Pearson, R., Brightmer, I. & Jegede, F. (1997) *Computer-aided assessment*, http://www.chelt.ac.uk/gdn/abstracts/a18.htm.

Perry, A. (1998) Netting research data for he climatologist, *Progress in Physical Geography*, 22(1), pp.121-126.

Pickles, J. (Ed.) (1995) *Ground Truth: the social implications of geographical information systems* (London: The Guildford Press).

Postman, N. (1985) *Amusing Ourselves to Death* (New York: Penguin).

Postman, N. (1995) *The End of Education* (New York: Knopf).

Proctor, J.D. & Richardson, A.E. (1997) Evaluating the effectiveness of multimedia computer modules as enrichment exercises for introductory human geography, *Journal of Geography in Higher Education*, 21(1), pp.41-55.

Purnell, K.N. & Solman, R.T. (1991) The influence of technical illustrations on student comprehension in geography, *Reading Research Quarterly*, 26(3), pp.277-299.

Raper, J.F. & Green, N.P.A. (1989) Development of a hypertext based tutor for geographical information systems, *British Journal of Educational Technology*, 3, pp.167-172.

Rees, P.H. (1987) Teaching computing skills to geography students, *Journal of Geography in Higher Education*, 11(2), 99-111.

Reeve, D. (1985) Computing in the geography degree: limitations and objectives, *Journal of Geography in Higher Education*, 9(1), 37-44.

Rich, D.C., Pitman, A.J., Gosper, M. & Jacobson, C. (1997) Restructuring of Australian higher education: information technology in geography teaching and learning, *Australian Geographer*, 28(2).

Riddle, D. (1990) Ecodisc CD-ROM, *The CTISS File*, Sept. No. 10, pp.27-31.

Rieber, L.P. (1994) *Computers, Graphics and Learning* (Madison: WI: Brown & Benchmark).

Ritter, M.E. (1997) *Earth Online: an Internet Guide for Earth Science* (Belmont, CA: Wadsworth Publishing Company).

Robinson, G. (1993) Use of computer-managed learning in quantitative methods course, *Journal of Geography in Higher Education*, 17(2), p.155.

Russell, T.L. (1995) *NB TeleEducation makes available the "No Significant Difference" phenomenon*, fourth edition, http://teleeducation.nb.ca/phenom/.

Schmeck, R. (Ed.) (1988) *Learning Strategies and Learning Styles* (New York: Plenum Press).

Shepherd, I.D.H. (1995) Multi-sensory GIS: mapping out the research frontier', in: T. Waugh (Ed.) *Proceedings of the 6th International Symposium on Spatial Data Handling*, pp.356-390 (London: Taylor & Francis).

Shepherd, I.D.H. (1996) The self-study approach to course delivery: developing a computer-supported geography module, *GeoCal*, 14, pp.6-9.

Shepherd, I.D.H. & Castleford, J. (1998) Evaluating the Web as an Educational Resource, forthcoming.

Stainfield, J. (1994) Use of a computer-based objective testing package in the assessment of geographical techniques, *Journal of Geography in Higher Education*, 18(2), pp.252-253.

Stainfield, J. (1997) Using IT to manage a third year module, *Active Learning*, 6, pp.30-34.

Stark, R. (1986) Demonstrating sociology: computers in the classroom, in: R. McGee (Ed.) *Teaching the mass class*, pp. 130-141 (American Sociological Association).

Starr, P. (1996) Computing our way to educational reform, *The American Prospect*, 27 (July-August), pp.50-60. (Also available on the Web at: http://epn.org/prospect/27/27star.html)

Stoll, C. (1995) *Silicon Snake Oil: second thoughts on the Information Superhighway* (New York: Doubleday).

Steeples, C. *et al.* (1998) International collaboration for geographic education on the World Wide Web, *Journal of Geography*, forthcoming.

Stoner, G. (Ed.) (1997) *Implementing Learning Technology*, http://www.icbl.hw.ac.uk/ltdi/implementing-it/cont.htm.

Taylor, L. (1997) Using the World Wide Web as a geography resource, *Teaching Geography*, 22(1), pp.11-15.

Tiffin, J. & Rajasingham, L. (1995) *In search of the virtual class: education in an information society* (London: Routledge).

Unwin, D.J. (1984) Things I do badly: tutorials, *Journal of Geography in Higher Education*, 8(2), pp.189-190.

Unwin, D.J. (1990) Using computers to help students learn: computer assisted learning in geography, *Area*, 23(1), pp.25-34.

Unwin, D.J. (1997) *Global climate change modelling*, http://www.chelt.ac.uk/gdn/abstracts/a32.htm.

Unwin, D.J. & Maguire, D. (1990) Developing the effective use of Information Technology in teaching and learning in geography: the Computers in Teaching Initiative Centre for Geography, *Journal of Geography in Higher Education*, 14(1), pp.77-82.

Unwin, D.J. & Wood, C.E. (Eds) (1990) *Readings in computer assisted learning in geography: a JGHE collection* (Leicester, University of Leicester: CTI Centre for Geography).

Veenema, S. & Gardner, H. (1996) Multimedia and multiple intelligences, *The American Prospect*, 27 (July-August), pp.69-75. (Also available on the Web at: http://epn.org/prospect/29/29veen.html).

Vincent, P., Chapman, G. & Steeples, C. (1997) *Tutorial groups and WWW conferencing*, http://www.chelt.ac.uk/gdn/abstracts/a54.htm.

Warburton, J. & Higgitt, M. (1997) Improving the preparation for fieldwork with IT: examples from physical geography, *Journal of Geography in Higher Education*, 21(3), pp.333-347.

Weaver, R. (1997) *Using optical mark readers in student assessment*, http://www.chelt.ac.uk/gdn/abstracts/a20.htm.

Webb, B. (1997) *'Paperless' lectures*, http://www.chelt.ac.uk/gdn/abstracts/a70.htm.

Webber, M.M. (1963) Order in diversity: community without propinquity, in: Lowdon Wingo (Ed.) *Cities and Space-The Future Use of Urban Land*, pp.25-54 (Baltimore: John Hopkins Press).

Weyer, S.A. & Borning, A. (1985) A prototype electronic encyclopedia, *Transactions on Office Information Systems*, 3(1), pp.62-88.

Whalley, B. (1997) *Critical use of the WWW and simple 'home page' construction*, http://www.chelt.ac.uk/gdn/abstracts/a66.htm.

Wilson, K.S. & Tally, W.J. (1991) *Designing for Discovery: interactive multimedia learning environments*, *Technical Report No. 15* (New York: Bank Street College of Education, Center for Technology in Education).

Wong, D. (1998) Creating a Web-based electronic reserve library for teaching world regional geography, *Journal of Geography in Higher Education*, 22(2), pp.259-262.

5.1.2   Other references

Boud, D., Keogh, R. & Walker, D. (1985) *Reflection: turning experience into learning* (London: Kogan Page).

CTI (Computers in Teaching Initiative) (1997) *Communication and Information Technologies for Teaching and Learning in Higher Education* (Oxford: CTI).

Dewey, J. (1944) *Democracy and Education* (New York: Macmillan).

Gamson, A.W. & Chickering, Z.F. (Eds) (1991) *Applying the seven principles for good practice in undergraduate education* (San Fransisco: Josey-Bass).

Hart, J.K. & Martinez, K. (1997) *Glacial Analysis: an interactive introduction* (London: Routledge).

Hassell, D. & Warner, H. (1995) *Using IT to Enhance Geography* (Sheffield: The Geographical Association; London: National Council for Educational Technology).

Healey, M., Robinson, G. & Castleford, J. (1997) Developing good educational practice: integrating *GeographyCAL* into university courses, in: *Proceedings of the Institute of Australian Geographers and New Zealand Geographical Society Second Joint Conference*, Department of Geography, The University of Waikato, Hobart, Australia.

Kemp, K.K. & Goodchild, F.M. (1992) Evaluating a major innovation in higher education: the NCGIA Core Curriculum in GIS, *Journal of Geography in Higher Education*, 16(1), pp.21-36.

Lave, J. (1988) *Cognition in Practice: mind, mathematics and culture in everyday life* (Cambridge: Cambridge University Press).

Lave, J. & Wenger, E. (1991) *Situated Learning: legitimate peripheral participation* (Cambridge: Cambridge University Press).

Livingstone, I. & Shepherd, I.D.H. (1997) Using the Internet, *Journal of Geography in Higher Education*, 21(3), pp.435-443.

Raper, J.R. & Green, N.P.A. (1989) GIST: an object-oriented approach to a GIS tutor, *Proceedings, Auto-Carto IX*, pp.610-619.

Schon, D. (1991) *Educating the Reflective Practitioner* (Oxford: Josey Bass).

Schupp, J.F. (1995) *Environmental Guide to the Internet* (Rockville, MD: Government Institutes Inc.).

Shepherd, I.D.H. (1991) Information integration and GIS, in: D. Maguire, M. F. Goodchild & D. F. Rhind (Eds) *Geographical Information Systems: principles and applications*, pp.337-360 (London: Longmans).

Silvert, W. (1984) Teaching ecological modelling with electronic spreadsheets, *Collegiate Microcomputer*, 2, 129-133.

5.2   References: the Internet and the Web

The following list of Web sites is not meant to be a 'Best of the Web' selection, but was compiled to illustrate key points made in this Guide. All documents included here were visited between November 1997 and February 1998, but there is no guarantee that they will still be available. Also, please note that there may be a painfully slow response from popular Web sites at peak visiting times.

[AAB]   Ask a biologist

              http://ls.la.asu.edu/askabiologist/index.html

[AAE]   Ask an expert

              http://www.askanexpert.com/askanexpert/index.shtml

[AAG]   Ask a geologist

              http://walrus.wr.usgs.gov/docs/ask-a-ge.html

[AAV]   Ask a Vulcanologist

              http://volcano.und.nodak.edu/

[AE]   AskERIC

              http://ericir.sunsite.syr.edu/

[AP]   The ACORN Project
         http://acorn.lboro.ac.uk/

[AR]   Architecture of Russia
         http://www.utexas.edu/depts/grg/virtdept/stylesheets/templates/
         format/slides/html_files/russia_toc.html

[ARI]   Aberystwyth resources on using the Internet
         http://www.aber.ac.uk/~ieswww/geores/geores.html

[ASCC]   Australian Schools curriculum communication scheme
         http://owl.qut.edu.au/oz-teachernet/

[BIDS]   Bath Information and Data Services (BIDS)
         http://www.bids.ac.uk/

[BOTH]   Gregory Bothun's evaluation of Web-based instruction
         http://zebu.uoregon.edu/tech.html

[BT]   BT corporate Web site
         http://www.bt.com/World/corpin

[BTCW]   BT's CampusWorld
         http://www.campus.bt.com/CampusWorld/

[BVF]   Bosnian Virtual Fieldtrip, George Mason University
         http://geog.gmu.edu/projects/bosnia/default.html

[CAQM]   Cambridge Air Quality Monitor
         http://www.io-ltd.co.uk/ccc.html

[CBT]   Computer-based tests 'white paper'
         http://www.qmark.com/company/1998paper.html

[CGF]   Critical Geography Forum
         http://www.mailbase.ac.uk/lists-a-e/crit-geog-forum

[CIS]   Computer Information Systems module at the US Military Academy
         http://www.eecs.usma.edu/cs383/cs383.htm

[CLUES] Centre for Land Use and Environmental Science
         http://www.clues.abdn.ac.uk:8080/directory/content.html

[CN]   ClydeNet
         http://cvu.strath.ac.uk/index.html

[COPAC] UK COPAC catalogue
         http://copac.ac.uk/copac/

[CP]   The CASTLE Project
         http://www.leicester.ac.uk/cc/ltg/castle/

[CTI]   Computers in Teaching Initiative
            http://www.cti.ac.uk/

[CTIG]   CTI Centre for Geography, Geology and Meteorology
            http://www.geog.le.ac.uk/cti/

[DL]   DeLiberations
            http://www.lgu.ac.uk/deliberations/

[DN]   Distributed.net
            http://www.distributed.net/

[DOOM]   Use of DOOM in Information Systems course
            http://www.eecs.usma.edu/cs383/doom/default.htm

[DSRS]   Dundee satellite receiving station
            http://www.sat.dundee.ac.uk/

[ECD]   Elsevier ContentsDirect
            http://www.elsevier.nl/locate/ContentsDirect

[ECN]   Econet
            http://www.igc.org/igc/econet/

[EDA]   ESRC Data Archive, University of Essex
            http://dawww.essex.ac.uk/

[EDC]   EROS Data Center
            http://edcwww.cr.usgs.gov/

[EDEM]   Index of DEMs on the Internet, Edinburgh University
            http://www.geo.ed.ac.uk/home/ded.html

[EDINA]   EDINA, Edinburgh University
            http://edina.ed.ac.uk/

[EDP]   Electronic Desktop Project Virtual Earthquake
            http://vflylab.calstatela.edu/edesktop/VirtApps/
            VirtualEarthQuake/VQuakeIntro.html

[EEL]   EE-Link
            http://www.nceet.snre.umich.edu/

[EEP]   The Electronic Emissary Project
            http://www.tapr.org/emissary/

[EOWD]   Environmental Organization WebDirectory
            http://www.webdirectory.com/

[EOWS]   The Earth Online Web site
            http://www.thomson.com/wadsworth/ritter/toc.html

[EPAMD]   EPA meteorological data

           http://www.epa.gov/scram001/t25.htm

[ERIC]   The Educational Resources Information Center (ERIC)

           http://www.aspensys.com/eric/

[EV]   The Electronic Volcano

           http://www.dartmouth.edu/~volcano/

[EVL]   Envirolink

           http://envirolink.org/index1.html

[EXG]   Excite geography pages

           http://www.excite.com:80/education/universities_and_colleges/
           fields_of_study/science/earth_sciences/geography

[FAO]   Food and Agriculture Organization

           http://apps.fao.org/

[FOE]   Friends of the Earth

           http://www.foe.org.uk/

[FTPSE]   FTP search engine

           http://ftpsearch.com/

[GDL]   Leicester Geography Department Web site

           http://www.le.ac.uk/CWIS/AD/GG/gg.html

[GEOSIM]   GeoSim Project

           http://geosim.cs.vt.edu/

[GFT]   Geology field trips on the Web

           http://www.uh.edu/~jbutler/anon/anontrips.html

[GIP]   Geo_Images Project

           http://geogweb.berkeley.edu/GeoImages.html

[GISCC]   NCGIA Core Curriculum in GIScience

           http://www.ncgia.ucsb.edu/education/curricula/giscc/welcome.html

[GL]   GlobaLearn

           http://www.globalearn.org/expeditions/brazil/index.html

[GLOBE]   Global Learning and Observations to Benefit the Environment Program

           http://globe.fsl.noaa.gov/

[GMU]   George Mason University geography pages

           http://geog.gmu.edu/gess/classes/default.html

[GN]   GeoNet

           http://www.pavilion.co.uk/dwakefield/

[GREEN]	Global Rivers Environmental Education Network
	http://www.igc.apc.org/green/
[GSNF]	Global schoolnet foundation
	http://www.gsn.org/
[GUBC]	Geoscience course at the University of British Columbia
	http://www.science.ubc.ca/~eoswr/
[HTM]	The Houston Traffic Monitor
	http://traffic.tamu.edu/traffic.html
[IFDT]	Interactive Frog Dissection Tutorial
	http://curry.edschool.Virginia.EDU/go/frog/
[ITCC]	Internet as a Tool for Collaborative Learning project
	http://www.lancs.ac.uk/users/edres/research/iheproject.html
[JOL]	JournalsOnline
	http://www.journalsonline.bids.ac.uk/Jol-pages/jol_body.html
[JP]	The JASON Project
	http://www.jasonproject.org/
[JSSM]	John Stainfield's simulation model
	http://www.science.plym.ac.uk/departments/geography/bihari.htm
[KB]	The Knowledge Base
	http://trochim.human.cornell.edu/kb/kbhome.htm
[KDRS]	Karl Donert's Remote Sensing pages
	http://www.livhope.ac.uk/livhope/ebs/ebswww/www/remote1.htm
[KE]	Kobe earthquake home page
	http://www.kobe-cufs.ac.jp/kobe-city/
[KTS]	Keirsey Temperament Sorter
	http://www.keirsey.com/cgi-bin/keirsey/newkts.cgi
[LAR]	Los Angeles Virtual River Tour
	http://www.lalc.k12.ca.us/target/units/river/riverweb.html
[LNL]	Use of Lotus Notes at Lancaster
	http://ktru-main.lancs.ac.uk/ISS/ltdolibr.nsf/
[LWT]	Lancaster University Web tutorial
	http://www.lancs.ac.uk/users/geogrphy/tutorial/tutorial.htm
	(note the deliberately missing 'a' in 'geogrphy')
[MAGG]	Magellan Internet Guide: geography reviews
	http://www.mckinley.com/magellan/Reviews/Science/Geography/index.magellan.html

[MB]   Mailbase

           http://www.mailbase.ac.uk/

[MEVC]   Mount Etna video camera

           http://www.iiv.ct.cnr.it:80/files/cam_index_etna.html

[MFVC]   Mount Fuji video camera (user controllable)

           http://www.flab.mag.keio.ac.jp/fuji/

           (Other Mount Fuji camera links can be found at http://dove.net.au/~punky/
           Fuji.html)

[MIDAS]   MIDAS, Manchester University

           http://midas.ac.uk/

[MNE]   MetNet Europe Schools Weather Project

           http://www.rmplc.co.uk/eduweb/sites/radgeog/MetNetEur/
           MetNetEur.html

[MUAWS]   Macquarie University Automatic Weather Station data

           http://atmos.es.mq.edu.au/aws/aws2/

[MV]   Montserrat volcano

           http://www.geo.mtu.edu/volcanoes/west.indies/soufriere/govt/

[NASAI]   NASA satellite imagery

           http://www.nasa.gov/NASA_homepage.html

[NASAKS]   NASA's KidSat site

           http://kidsat.jpl.nasa.gov/

[NCDC]   National climatic data center

           http://www.ncdc.noaa.gov/

[NCES]   Nene College Environmental Science

           http://www.nene.ac.uk/aps/env/starter.html

[NGS]   National Geographic Society

           http://www.NationalGeographic.com/main.html

[NGSS]   National geographic society's survey 2000

           http://survey2000.nationalgeographic.com/

[NISS]   NISS Information Gateway

           http://www.niss.ac.uk

[NL]   NetLearn

           http://www.rgu.ac.uk/~sim/research/netlearn/callist.htm

[NS]   Netskills

           http://www.netskills.ac.uk

[NSV]   NASA's SkyView Virtual Observatory
http://skyview.gsfc.nasa.gov/

[OITT]   Oxford Brookes University IT Term
http://www.brookes.ac.uk/it_term/

[OSMV]   One sky, many voices project
http://onesky.engin.umich.edu/

[OUVM]   Open University's Virtual Microscope
http://met.open.ac.uk/vms/vms.html

[OWO]   Oneworld Online
http://www.oneworld.org/

[PAAO]   Project atmosphere Australia online
http://www.schools.ash.org.au/paa/paa.htm

[RA]   Research assistance
http://www.research-assistance.com/cgi-bin/hazel-cgi/hazel.cgi

[RAN]   Rainforest Action Network
http://www.ran.org/ran/

[RSCC]   Remote Sensing Core Curriculum
http://research.umbc.edu/~tbenja1/

[SETI]   SETI institute - SETI@home project
http://setiathome.ssl.berkeley.edu/

[SL]   StatLib
http://www.stat.cmu.edu/

[SOSIG]   Social Science Information Gateway
http://www.sosig.ac.uk/

[SPGSGS] St Paul's Grammar School Geography Jumpsite
http://apostle.stpauls.nsw.edu.au/IBGEO.html

[STM]   Seattle Traffic Map
http://www.wsdot.wa.gov/regions/northwest/nwflow/

[TGC]   The Geographer's Craft
http://www.utexas.edu/depts/grg/gcraft/contents.html

[TLS]   The Learning Station (BBC)
http://www.bbc.co.uk/education/schools/

[TLTPC]   TLTP on-line catalogue
http://www.niss.ac.uk/education/cti/cti.html

[TNA]   The Networking Academy

            http://www.lancs.ac.uk/users/edres/research/NetAcad/home.html

[TQ]   TerraQuest

            http://www.terraquest.com/

[TRIADS]   TRIADS

            http://www.pcweb.liv.ac.uk/apboyle/triads/index.html

[TSP]   The seven principles of good practice

            http://www.tltgroup.org/ehrmann.htm

[UCES]   University of California at Santa Cruz Earth Sciences course

            http://wwwcatsic.ucsc.edu/~eart8/

            (Other courses from the Earth Sciences department are at
            http://emerald.ucsc.edu/education/escourses/)

[ULW]   USGS Learning Web site

            http://www.usgs.gov/education/

[UNMI]   University of Nottingham meteosat images

            http://www.nottingham.ac.uk/meteosat/

[UPCM]   University of Portsmouth geography course materials

            http://www.envf.port.ac.uk/geog/teaching/environ/echome.htm

[USCB]   US Census Bureau

            http://www.census.gov/

[USDA]   US Department of Agriculture

            http://hydrolab.arsusda.gov/

[USE]   USGS Earthquakes site

            http://quake.wr.usgs.gov/

[VE]   Volcanic eruptions

            http://volcano.und.nodak.edu/

[VFC]   The Virtual Fieldcourse

            http://www.geog.le.ac.uk/vfc/

[VFCS]   Virtual Fieldcourse sites

            http://www.geog.le.ac.uk/cti/virt.html

[VFDK]   Virtual Frog Dissection Kit Lawrence Berkeley National Laboratory

            http://www-itg.lbl.gov/ITG.hm.pg.docs/dissect/dissect.html

[VFMA]   Aberystwyth Virtual Fieldcourse to Malta

            http://www.aber.ac.uk/~jpg/malta/maltind.html

[VFMP]   Portsmouth Virtual Fieldcourse to Malta

            http://sh.plym.ac.uk/TeachingLearning/pb_malta.htm

[VG]   Virtual Galapagos

            http://www.terraquest.com/galapagos/

[VGD]   Virtual Geography Department, University of Texas

            http://www.utexas.edu/depts/grg/virtdept/contents.html

            http://www.utexas.edu/depts/grg/virtdept/about/about.html

[VLP]   World Wide Web Consortium Virtual Library Project

            http://www.tissot.org/vl/coordination.html

[VLPE]   World Wide Web Consortium Virtual Library Project: Environment

            http://earthsystems.org/Environment.shtml

[VLPG]   World Wide Web Consortium Virtual Library Project: Geography

            http://www.icomos.org/Geography/WWW_VL_Geography.html

[VOU]   The Virtual Online University

            http://www.athena.edu

[WCS]   Web cam sites

            http://telusplanet.net/public/imco/cams.htm

[WLHG]   World Lecture Hall: Geography

            http://wwwhost.cc.utexas.edu/world/lecture/geography/

[WW]   Weatherworks weather links

            http://www.weatherworks.com/links.html

[YAG]   Yahoo geography pages

            http://www.yahoo.com/Science/Geography/

[YWL]   Yahoo index of world locations

            http://www.yahoo.com/regional/countries/

5.3   Resources

The items listed below may be of some use in following up some of the ideas presented in this Guide. Many of these are included in the Resources Database, so you will find them when searching for particular topics or by entering selected keywords.

ICT Resources

Geographical courseware and other computer resources can be tracked down in a number of places. Academic sources include:

GeographyCal Project. A dozen or so courseware modules are now available, and others are in production. The software can be acquired from the Bath University Computing Service [BUCS]. Several other subject-based TLTP projects produce materials relevant to geography teaching; for example, the Centre for

Land Use and Environmental Science [CLUES]. A searchable catalogue of all current TLTP courseware is available [TLTPC], and a printed catalogue is also available.

Individual institutions. For example, the GeoSim project at the Virginia Tech has made its simulation models available on the Internet [GEOSIM], while staff at the University of Southampton have teamed up with commercial publishers to release Glacial Analysis (Hart & Martinez 1997).

A useful directory of Internet resources for teaching and learning is provided by Netlearn [NL].

Commercial sources include:

- Encarta (from Microsoft)

Non-profit-making sources include:

- GeoMedia (USGS)
- GREEN environmental monitoring kit [GREEN]

Resource Reviews

Reviews of educational IT resources appear in:

- *Active Learning* (CTISS publication)
- CTI Newsletter
- *GeoCAL* (CTICG publication)
- *Journal of Geography in Higher Education*
- *Professional Geographer*
- *Teaching Geography*
- TLTP/CTI Newsletters

Information

GeoCAL The newsletter of the CTI Centre for Geography Geology and Meteorology, which is published on a regular basis from the Department of Geography at Leicester University. It provides regular reports on recently released courseware in geography, many examples of which are reviewed, and it lists useful Web sites.

The CTI Centre for Geography, Geology and Meteorology Web site is based at the University of Leicester, and provides a wealth of useful links to on-line resources for teaching and learning geography [CTIG].

A free journal alert service is provided by many journal publishers (for example, to register at Elsevier's ContentsDirect, send an email to cdsubs@elsevier.co.uk or visit its Web site [ECD].

Other support materials

A variety of non-computer materials are becoming available (for example, tutorials, lectures, technology briefings) which help teachers adopt ICT for their teaching. The Web provides an increasing amount of such materials for geography teaching and learning. Examples include:

> The Geographer's Craft (University of Texas) [TGC]

> The Virtual Geography Department (University of Texas) [VGD]

A number of departments have developed introductory guides to using the Internet and the Web for their students, which include links to relevant Web sites. See, for example, the 'starter pack' at Nene College for Environmental Science students [NCES], the Internet resources page at Aberystwyth for Earth Studies students [ARI], and the tutorial on Web resources at Lancaster for Geography students (which includes a glossary of Internet terminology) [LWT]. Perhaps the best introductory tutorial on using the Internet and the Web (TONIC) is provided by the TLTP-funded Netskills Project [NS].

Reports of user experiences

A series of reports on the educational use of IT in UK geography departments has appeared over the past several issues of *GeoCAL*. The *Journal of Geography in Higher Education* publishes extended reports of ICT-supported geography teaching. (A complete index is available in the Resources Database.) Case studies of the use of IT to enhance geography teaching in the schools sector can be found in Hassell & Warner (1995).

Discussion

Several 'discussion lists' are available on the Internet that can be used by geography teachers to discuss educational issues. The moderated GeogNet discussion list based at Nene College is the premier UK venue for such electronic exchanges. To subscribe, simply send an empty email message to: GeogNet@nene.ac.uk. Other discussion forums which feature geography and teaching issues include DeLiberations [DL] and crit-geog-forum [CGF]. You can find other discussion lists relevant to teaching and learning by using the Mailbase service [MB].

5.4   Glossary

The worlds of IT and educational technology are replete with acronyms and technical jargon. Although we have tried to avoid using most of this vocabulary in this Guide, some of it is essential. The following glossary is meant for those who may be new to this particular field, but feel it worth persevering despite the off-putting language often used by many of its practitioners.

Asynchronous	Refers to communication where interactions are not in real time (for example, mail, email, newsgroups).
Broadband	Refers to networks which are able to carry large amounts of information simultaneously.
CAL	Computer-assisted learning: the broad use of computers to support and enhance teaching and learning.

CBL	Computer-based learning; equivalent to CAL.
CEL	Computer-enhanced learning; equivalent to CAL.
CSL	Computer-supported learning; equivalent to CAL.
C&IT	Communications and Information Technology, a phrase introduced by the Dearing Report (NCIHE 1997) to broaden thinking about technology in education. Its definition reads: "Those technologies which enable the processing, storage and transmission of both live and recorded information by electronic means." Also used by the CTI (1997). (See also ICT.)
CML	Computer-managed learning: the use of computers to manage the learning process.
CT	Communication technology: originally analogue, but increasingly digital, technologies that are used for communicating between people and machines (especially computers).
CTI	Computers in Teaching Initiative: a national project, initiated in 1984, to encourage the use of computers in teaching and learning in various disciplines in higher education in the UK (CTI Support Service email: CTISS@oucs.ox.ac.uk CTI Web address: [CTI]).
CTIGGM	The CTI disciplinary centre for Geography, Geology and Meteorology, located in the Department of Geography at the University of Leicester (email: ctigeog@le.ac.uk). It provides a major Web site for educational uses of information and communication technologies [CTIG].
Courseware	Computer software used in (and usually designed for) teaching and learning.
Email	Common abbreviation for 'electronic mail', which refers to the digital messages sent from one person's computer to another. Although most email messages consist of text, email software usually allows non-textual material to be 'attached' (for example, word-processed documents, graphical images, video clips, sounds). Like FAX, email allows messages to be sent and received almost instantly — hence the disparaging term 'snail mail' used to refer to conventional postal services. Unlike conventional FAX, however, email messages can be stored on the receiving person's computer, or on a mail-handling computer (or 'mailserver'), until the receiver is ready to read their mail.
FAQ	Common abbreviation for a list of answers to the most 'Frequently Asked Questions' on a particular topic. Many such lists appear on Web sites.
File Transfer	The practice of sending information from one computer to another, traditionally on floppy disks and CD-ROMs, but increasingly across private computer networks or the Internet. On the Internet, there is a standard set of rules, known as the 'File Transfer Protocol' (or ftp) that is obeyed by computers wishing to exchange files with minimum user intervention.

GIF The standard file format for graphical images on the Web (Graphics Interchange Format).

GeogNet The discussion list on the Internet for geographers interested in educational issues, based at Nene College (Internet address: GeogNet@nene.ac.uk).

GeographyCal The geographical courseware developed under the TLTP Phase 2 project at the CTI Centre for Geography, University of Leicester.

GIT Geographical information technology, a catch-all phrase for technologies that have significant geographical applications.

GPS Global Positioning System. The system of 24 satellites placed in earth orbit by the US to provide worldwide locational fixes for the military. Now widely used by the civilian sector for mapping, navigation and educational applications.

Groupware Software that facilitates people working together in groups.

HTML Hypertext markup language. The set of tagging conventions used to create text documents for exchange over the World Wide Web.

HTTP The hypertext transfer protocol. The set of rules governing the exchange of hypertext (i.e. HTML) documents over the World Wide Web.

Hypermedia Multimedia materials that have a non-linear structure, due to the embedding of links in the information. The Web is a global example of hypermedia.

ICT Information and Communication Technologies, a term used throughout this Guide to refer to the rapidly converging technologies of information processing and communications. Also used in HM Government's National Grid for Learning proposals published in 1997, by the National Council for Educational Technology, and the European Commission, among others. (See also C&IT.)

Internet Shorthand for the global network of computers and computer networks that has developed primarily since the 1980s, over which people can communicate and share information digitally.

Intranet A local version of the Internet that is internal to a particular organisation.

IT Information Technology: the technologies (especially computer-based) developed to handle digital information.

JPEG The standard file format for photographs on the Web (developed by the Joint Photographic Experts Group).

Multimedia Digital information in various forms, including text, numbers, graphics, video and sound.

NCGIA The National Center for Geographical Information Analysis. The national centre for research and development in GIS (including educational issues) in the USA.

NDPCAL	The National Development Programme for Computer Assisted Learning, the first nationally funded programme to develop educational software in the UK in the mid-1970s.
Synchronous	Refers to communication that involves real-time interaction (for example, telephone, video-conferencing, chat systems).
Telematics	A combination of (digital) computer and communication technologies.
TLTP	The Teaching and Learning Technology Programme, set up in 1992, and currently based at the University of Bath [TLTP]. Now entering its third phase of funding, this national project provides financial support for the creation of courseware in a wide variety of disciplines in higher education, including geography.
URL	Abbreviation for Uniform Resource Locator, the 'address' of a document on the Web.

World Wide Web, WWW or Web

The 'network' of specially formatted documents that are available on computers connected to the Internet. These documents typically consist of text, but they increasingly incorporate graphical images, animations and other information which can be displayed by modern Web 'browsers' (the name for the software that is used to explore Web information).

5.5   Acknowledgements

To John Castleford and Geoff Robinson at the CTI Centre for Geography, Geology and Meteorology at the University of Leicester, for invaluable help in tracking down examples of good practice in the use of IT in geography, and for permission to include a revised version of the annotated CAL Bibliography in the Resources Database. To David Riley, University of North London, for permission to include a version of his cumulative index to the *Journal of Geography in Higher Education* in the Resources Database. And to the many other geographers and teachers who have contributed ideas and case studies, and provided critical comments on earlier versions of this Guide.